The Executive Guide to Artificial Intelligence

Andrew Burgess

The Executive Guide to Artificial Intelligence

How to identify and implement applications for AI in your organization

palgrave
macmillan

Andrew Burgess
AJBurgess Ltd
London, United Kingdom

ISBN 978-3-319-63819-5   ISBN 978-3-319-63820-1   (eBook)
https://doi.org/10.1007/978-3-319-63820-1

Library of Congress Control Number: 2017955043

Cover illustration: Ukususha/iStock/Getty Images Plus

Printed on acid-free paper

This Palgrave Macmillan imprint is published by Springer Nature
The registered company is Springer International Publishing AG
The registered company address is: Gewerbestrasse 11, 6330 Cham, Switzerland

This book is dedicated to my wonderful wife, Meg, and our two amazing children, James and Charlie.

Foreword

I remember well the AI work I did whilst at college studying computer science, how different and fascinating it was and still is. We were set a very open challenge to write an AI programme on any subject. I decided to write mine so that it could tell you if the building in a photo was a house, a flat or a bungalow. Somewhat impractical, but a great learning experience for me, particularly in understanding how AI is different from traditional software.

Although my college days were a number of years ago, since that time the concept of computers learning has always intrigued me and I have since wondered how long it will take for AI to have a truly widespread impact. In recent years, we've seen massive improvements in processing power, big data collection via sensors and the Internet of Things, cloud services, storage, ubiquitous connectivity and much more. These technological leaps mean that this is the right time for AI to become part of the 'here and now' and I strongly believe we will see a dramatic increase in the use of AI over the next few years.

The AI in use today is known as narrow AI because it can excel at thousands of relatively narrow tasks (e.g. doing internet searches, playing Go or looking for fraudulent transactions). Things will certainly get even more exciting when 'general AI' can outperform humans at nearly every task we do, but we simply don't know when this might be, or what the world will then look like. Until then, what excites me most is how we can apply AI now to solve our day-to-day problems at home and work.

So why is AI important and how can we use it? Firstly, if you are impatient (like I am), doing small manual, repetitive tasks on computers simply takes too much time. I want the computer to do a better job of anticipating my needs and to just get on with it. If I could, I would prefer to talk to Alexa or

Google Assistant and just tell the computer what to do. For example, I would love to be able to ask Alexa to buy the most convenient train ticket and take the money out of my account. Compare this to buying a train ticket on any website, where after something like 50 key strokes you might have bought a ticket. I don't think future generations, who are becoming increasingly impatient, will put up with doing these simple and time-consuming tasks. I see my children and future generations having more 'thinking time' and focusing on things that are outside the normal tasks. AI may in fact free up so much of my children's time that they can finally clean up their bedrooms.

In the workplace, how many of the emails, phone calls and letters in a call centre could be handled by AI? At Virgin Trains, we used AI to reduce the time spent dealing with customer emails by 85% and this enabled our people to focus on the personable customer service we're famous for. Further improvements will no doubt be possible in the future as we get better at developing conversational interfaces, deep learning and process automation. One can imagine similar developments revolutionising every part of the business, from how we hire people to how we measure the effectiveness of marketing campaigns.

So, what about the challenges of AI? One that springs to mind at Virgin is how to get the 'tone of voice' right. Our people are bold, funny and empathetic, and our customers expect this from us in every channel. Conversational interfaces driven by AI should be no different.

Today it may be a nuisance if your laptop crashes, but it becomes all the more important that an AI system does what you want it to do if it controls your car, your airplane or your pacemaker. With software systems that can learn and adapt, we need to understand where the responsibility lies when they go wrong. This is both a technical and an ethical challenge. Beyond this, there are questions about data privacy, autonomous weapons, the 'echo chamber' problem of personalised news, the impact on society as increasing numbers of jobs can be automated and so on.

Despite these challenges, I am incredibly excited about the future of technology, and AI is right at the heart of the 'revolution'. I think over the next five to ten years AI will make us more productive at work, make us more healthy and happy at home, and generally change the world for the better.

To exploit these opportunities to the full, businesses need people who understand these emerging technologies and can navigate around the challenges. This book is essential reading if you want to understand this transformational technology and how it will impact your business.

John Sullivan, CIO and Innovation at Virgin Trains

Acknowledgements

I would like to thank the following people for providing valuable input, content and inspiration for this book:

Andrew Anderson, Celaton
Richard Benjamins, Axa
Matt Buskell, Rainbird
Ed Challis, Re:infer
Karl Chapman, Riverview Law
Tara Chittenden, The Law Society
Sarah Clayton, Kisaco Research
Dana Cuffe, Aldermore
Rob Divall, Aldermore
Gerard Frith, Matter
Chris Gayner, Genfour
Katie Gibbs, Aigen
Daniel Hulme, Satalia
Prof. Mary Lacity, University of Missouri-St Louis
Prof. Ilan Oshri, Loughborough University
Stephen Partridge, Palgrave
Mike Peters, Samara
Chris Popple, Lloyds Bank
John Sullivan, Virgin Trains
Cathy Tornbaum, Gartner

Vasilis Tsolis, Congnitiv+
Will Venters, LSE
Kim Vigilia, Conde Naste
Prof. Leslie Willcocks, LSE
Everyone at Symphony Ventures

Contents

List of Figures

1

Don't Believe the Hype

Introduction

Read any current affairs newspaper, magazine or journal, and you are likely to find an article on artificial intelligence (AI), usually decrying the way the 'robots are taking over' and how this mysterious technology is the biggest risk to humanity since the nuclear bomb was invented. Meanwhile the companies actually creating AI applications make grand claims for their technology, explaining how it will change peoples' lives whilst obfuscating any real value in a mist of marketing hyperbole. And then there is the actual technology itself—a chimera of mathematics, data and computers—that appears to be a black art to anyone outside of the developer world. No wonder that business executives are confused about what AI can do for their business. What exactly is AI? What does it do? How will it benefit my business? Where do I start? All of these are valid questions that have been, to date, unanswered, and which this book seeks to directly address.

Artificial Intelligence, in its broadest sense, will have a fundamental impact on the way that we do business. Of that there is no doubt. It will change the way that we make decisions, it will enable completely new business models to be created and it will allow us to do things that we never before thought possible. But it will also replace the work currently being done by many knowledge workers, and will disproportionally reward those who adopt AI early and effectively. It is both a huge opportunity and an ominous threat wrapped up in a bewildering bundle of algorithms and jargon.

But this technological revolution is not something that is going to happen in the future; this is not some theoretical exercise that will concern a few businesses.

© The Author(s) 2018
A. Burgess, *The Executive Guide to Artificial Intelligence*,
https://doi.org/10.1007/978-3-319-63820-1_1

Artificial Intelligence is being used today in businesses to augment, improve and change the way that they work. Enlightened executives are already working out how AI can add value to their businesses, seeking to understand all the different types of AI and working out how to mitigate the risks that it inevitably brings. Many of those efforts are hidden or kept secret by their instigators, either because they don't want the use of AI in their products or services to be widely known, or because they don't want to give away the competitive advantage that it bestows. A persistent challenge for executives that want to get to grips with AI is where to find all the relevant information without resorting to fanciful articles, listening to vendor hyperbole or trying to understand algorithms. AI is firmly in the arena of 'conscious unknowns'—we know that we don't know enough.

People generally experience AI first as consumers. All our smartphones have access to sophisticated AI, whether that is Siri, Cortana or Google's Assistant. Our homes are now AI enabled through Amazon's Alexa and Google Home. All of these supposedly make our lives easier to organise, and generally they do a pretty good job of it. But their use of AI is actually pretty limited. Most of them rely on the ability to turn your speech into words, and then those words into meaning. Once the intent has been established, the rest of the task is pretty standard automation; find out the weather forecast, get train times, play a song. And, although the speech recognition and natural language understanding (NLU) capabilities are very clever in what they achieve, AI is so much more than that, especially in the world of business.

Artificial Intelligence can read thousands of legal contracts in minutes and extract all the useful information out of them; it can identify cancerous tumours with greater accuracy than human radiologists; it can identify fraudulent credit card behaviour before it happens; it can drive cars without drivers; it can run data centres more efficiently than humans; it can predict when customers (and employees) are going to desert you and, most importantly, it can learn and evolve based on its own experiences.

But, until business executives understand what AI is, in simple-enough terms, and how it can help their business, it will never reach its full potential. Those with the foresight to use and exploit AI technologies are the ones that need to know what it can do, and understand what they need to do to get things going. That is the mission of this book. I will, over the course of the ten chapters, set out a framework to help the reader get to grips with the eight core capabilities of AI, and relate real business examples to each of these. I will provide approaches, methodologies and tools so that you can start your AI journey in the most efficient and effective way. I will also draw upon interviews and case studies from business leaders who are already implementing AI, from established AI vendors, and from academics whose work focuses on the practical application of AI.

Introducing the AI Framework

My AI Framework was developed over the past few years through a need to be able to make sense of the plethora of information, misinformation and marketing-speak that is written and talked about in AI. I am not a computer coder or an AI developer, so I needed to put the world of AI into a language that business people like myself could understand. I was continually frustrated by the laziness in the use of quite specific terminology in articles that were actually meant to help explain AI, and which only made people more confused than they were before. Terms like Artificial Intelligence, Cognitive Automation and Machine Learning were being used interchangeably, despite them being quite different things.

Through my work as a management consultant creating automation strategies for businesses, through reading many papers on the subject and speaking to other practitioners and experts, I managed to boil all the available information down into eight core capabilities for AI: Image Recognition, Speech Recognition, Search, Clustering, NLU, Optimisation, Prediction and Understanding. In theory, any AI application can be associated with one or more of these capabilities.

The first four of these are all to do with **capturing information**—getting structured data out of unstructured, or big, data. These Capture categories are the most mature today. There are many examples of each of these in use today: we encounter Speech Recognition when we call up automated response lines; we have Image Recognition automatically categorising our photographs; we have a Search capability read and categorise the emails we send complaining about our train being late and we are categorised into like-minded groups every time we buy something from an online retailer. AI efficiently captures all this unstructured and big data that we give it and turns it into something useful (or intrusive, depending on your point of view, but that's a topic to be discussed in more detail later in the book).

The second group of NLU, Optimisation and Prediction are all trying to work out, usually using that useful information that has just been captured, **what is happening**. They are slightly less mature but all still have applications in our daily lives. NLU turns that speech recognition data into something useful—that is, what do all those individual words actually mean when they are put together in a sentence? The Optimisation capability (which includes problem solving and planning as core elements) covers a wide range of uses, including working out what the best route is between your home and work. And then the Prediction capability tries to work out what will happen next— if we bought that book on early Japanese cinema then we are likely to want to buy this other book on Akira Kurosawa.

Once we get to Understanding, it's a different picture all together. Understanding **why something is happening** really requires cognition; it requires many inputs, the ability to draw on many experiences, and to conceptualise these into models that can be applied to different scenarios and uses, which is something that the human brain is extremely good at, but AI, to date, simply can't do. All of the previous examples of AI capabilities have been very specific (these are usually termed Narrow AI) but Understanding requires *general* AI, and this simply doesn't exist yet outside of our brains. Artificial General Intelligence, as it is known, is the holy grail of AI researchers but it is still very theoretical at this stage. I will discuss the future of AI in the concluding chapter, but this book, as a practical guide to AI in business today, will inherently focus on those Narrow AI capabilities that can be implemented now.

You will already be starting to realise from some of the examples I have given already that when AI is used in business it is usually implemented as a combination of these individual capabilities strung together. Once the individual capabilities are understood, they can be combined to create meaningful solutions to business problems and challenges. For example, I could ring up a bank to ask for a loan: I could end up speaking to a machine rather than a human, in which case AI will first be turning my voice into individual words (Speech Recognition), working out what it is I want (NLU), deciding whether I can get the loan (Optimisation) and then asking me whether I wanted to know more about car insurance because people like me tend to need loans to buy cars (Clustering and Prediction). That's a fairly involved process that draws on key AI capabilities, and one that doesn't have to involve a human being at all. The customer gets great service (the service is available day and night, the phone is answered straight away and they get an immediate response to their query), the process is efficient and effective for the business (operating costs are low, the decision making is consistent) and revenue is potentially increased (cross-selling additional products). So, the combining of the individual capabilities will be key to extracting the maximum value from AI.

The AI Framework therefore gives us a foundation to help understand what AI can do (and to cut through that marketing hype), but also to help us apply it to real business challenges. With this knowledge, we will be able to answer questions such as; How will AI help me enhance customer service? How will it make my business processes more efficient? And, how will it help me make better decisions? All of these are valid questions that AI can help answer, and ones that I will explore in detail in the course of this book.

Defining AI

It's interesting that in most of the examples I have given so far people often don't even realise they are actually dealing with AI. Some of the uses today, such as planning a route in our satnav or getting a phrase translated in our browser, are so ubiquitous that we forget that there is actually some really clever stuff happening in the background. This has given rise to some tongue-in-cheek definitions of what AI is: some say it is anything that will happen in 20 years' time, others that it is only AI when it looks like it does in the movies. But, for a book on AI, we do need a concise definition to work from.

The most useful definition of AI I have found is, unsurprisingly, from the *Oxford English Dictionary*, which states that AI is "the theory and development of computer systems able to perform tasks normally requiring human intelligence". This definition is a little bit circular since it includes the word 'intelligence', and that just raises the question of what is intelligence, but we won't be going into that philosophical debate here.

Another definition of AI which can be quite useful is from Andrew Ng, who was most recently the head of AI at the Chinese social media firm, Baidu, and a bit of a rock star in the world of AI. He reckons that any cognitive process that takes a human under one second to process is a potential candidate for AI. Now, as the technologies get better and better this number may increase over time, but for now it gives us a useful benchmark for the capabilities of AI.

Another way to look at AI goes back to the very beginnings of the technology and a very fundamental question: should these very clever technologies seek to *replace* the work that human beings are doing or should they *augment* it? There is a famous story of the two 'founders' of AI, both of whom were at MIT: Marvin Minsky and Douglas Engelbart. Minsky declared "We're going to make machines intelligent. We are going to make them conscious!" To which Engelbart reportedly replied: "You're going to do all that for the machines? What are you going to do for the people?" This debate is still raging on today, and is responsible for some of those 'robots will take over the world' headlines that I discussed at the top of this chapter.

The Impact of AI on Jobs

It is clear that AI, as part of the wider automation movement, will have a severe impact on jobs. There are AI applications, such as chatbots, which can be seen as direct replacements for call centre workers. The ability to read thousands of documents in seconds and extract all the meaningful information from them will hollow out a large part of the work done by accountants and junior lawyers. But equally, AI can augment the work done by these groups as well. In the call centre, cognitive reasoning systems can provide instant and intuitive access to all of the knowledge that they require to do their jobs, even if it is their first day on the job—the human agent can focus on dealing with the customer on an emotional level whilst the required knowledge is provided by the AI. The accountants and junior lawyers will now have the time to properly analyse the information that the AI has delivered to them rather than spend hours and hours collating data and researching cases.

Whether the net impact on work will be positive or negative, that is, will automation create more jobs than it destroys, is a matter of some debate. When we look back at the 'computer revolution' of the late twentieth century that was meant to herald massive increases in productivity and associated job losses, we now know that the productivity benefits weren't as great as people predicted (PCs were harder to use than first imagined) and the computers actually generated whole new industries themselves, from computer games to movie streaming. And, just like the robots of today, computers still need to be designed, manufactured, marketed, sold, maintained, regulated, fixed, fuelled, upgraded and disposed of.

The big question, of course, is whether the gains from associated activities plus any new activities created from automation will outweigh the loss of jobs that have been replaced. I'm an optimist at heart, and my own view is that we will be able to adapt to this new work eventually, but not before going through a painful transition period (which is where a Universal Basic Income may become a useful solution). The key factor therefore is the pace of change, with all the indicators at the moment suggesting that the rate will only get faster in the coming years. It's clear that automation in general, and AI in particular, are going to be huge disruptors to all aspects of our lives—most of it will be good but there will be stuff that really challenges our morals and ethics. As this book is a practical guide to implementing AI now, I'll be exploring these questions in a little more detail toward the end of the book, but the main focus is very much on the benefits and challenges of implementing AI today.

A Technology Overview

The technology behind AI is fiendishly clever. At its heart, there are algorithms: an algorithm is just a sequence of instructions or a set of rules that are followed to complete a task, so it could simply be a recipe or a railway timetable. The algorithms that power AI are essentially very complicated statistical models—these models use probability to help find the best output from a range of inputs, sometimes with a specific goal attached ('if a customer has watched these films, what other films would they also probably like to watch?'). This book is certainly not about explaining the underlying AI technology; in fact, it is deliberately void of technology jargon, but it is worth explaining some of the principles that underpin the technology.

One of the ways that AI technologies are categorised is between 'supervised' and 'unsupervised' learning. Supervised learning is the most common approach out of the two and refers to situations where the AI system is trained using large amounts of data. For example, if you wanted to have an AI that could identify pictures of dogs then you would show it thousands of pictures, some of which had dogs and some of which didn't. Crucially, all the pictures would have been labelled as 'a dog picture' or 'not a dog picture'. Using machine learning (an AI technique which I'll come on to later) and all the training data the system learns the inherent characteristics of what a dog looks like. It can then be tested on another set of similar data which has also been tagged but this time the tags haven't been revealed to the system. If it has been trained well enough, the system will be able to identify the dogs in the pictures, and also correctly identify pictures where there is no dog. It can then be let loose on real examples. And, if the people using your new 'Is There a Dog in My Picture?' app are able to feed back when it gets it right or not, then the system will continue to learn as it is being used. Supervised learning is generally used where the input data is unstructured or semi-structured, such as images, sounds and the written word (Image Recognition, Speech Recognition and Search capabilities in my AI Framework).

With Unsupervised Learning, the system starts with just a very large data set that will mean nothing to it. What the AI is able to do though is to spot clusters of similar points within the data. The AI is blissfully naive about what the data means; all it does is look for patterns in vast quantities of numbers. The great thing about this approach is that the user can also be naive—they don't need to know what they are looking for or what the connections are—all that work is done by the AI. Once the clusters have been identified then predictions can be made for new inputs.

So, as an example, we may want to be able to work out the value of a house in a particular neighbourhood. The price of a house is dependent on many variables such as location, number of rooms, number of bathrooms, age, size of garden and so on, all of which make it difficult to predict its value. But, surely there must be some complicated connection between all of these variables, if only we could work it out? And that's exactly what the AI does. If it is fed enough base data, with each of those variables as well as the actual price, then it uses statistical analysis to find all the connections—some variables may be very strong influencers on price whilst others may be completely irrelevant. You can then input the same variables for a house where the price is unknown and it will be able to predict that value. The data that is input is structured data this time, but the model that is created is really a black box. This apparent lack of transparency is one of AI's Achilles's heels, but one that can be managed, and which I'll discuss later in the book.

As well as the above two types of training, there are various other terms associated with AI, and which I'll cover briefly here, although for business executives they only need to be understood at a superficial level. 'Neural Networks' is the term used to describe the general approach where AI is mimicking the way that the brain processes information—many 'neurons' (100 billion in the case of the brain) are connected to each other in various degrees of strength, and the strength of the connection can vary as the brain/machine learns.

To give an over-simplified example, in the dog picture app above, the 'black nose' neuron will have a strong influence on the 'dog' neuron, whereas a 'horn' neuron will not. All of these artificial neurons are connected together in layers, where each layer extracts an increasing level of complexity. This gives rise to the term Deep Neural Networks. Machine Learning, where the machine creates the model itself rather than a human creating the code (as in the examples I have given above), uses DNNs. So, think of these terms as concentric circles: AI is the over-arching technology, of which Machine Learning is a core approach that is enabled by DNNs.

There are obviously many more terms that are in common use in the AI world, including decision tree learning, inductive logic programming, reinforcement learning and Bayesian networks, but I will cover these only when absolutely necessary. The focus of this book, as you will now hopefully appreciate, is on the business application of AI rather than its underlying technologies.

About This Book

My working experience has been as a management consultant, helping organisations cope with the challenges of the time, from productivity improvement, through change management and transformation to outsourcing and robotic process automation, and now AI. I first came across AI properly in my work in 2001. I was working as Chief Technology Officer in the Corporate Venturing unit of a global insurance provider—my role was to identify new technologies that we could invest in and bring into the firm as a foundation client (it was what we used to call the 'incubator' model). One of the technologies we invested in was based around the idea of 'smart agents' that could be used as an optimisation engine—each agent would have a specific goal and would 'negotiate' with the other agents to determine the ideal collective outcome. So, for example, the system could determine the most effective way for trucks to pass through a port, or the best way to generate the most revenue from the size and arrangement of advertisements in newspapers. Although we didn't call it AI at the time, this is effectively what it was—using computer algorithms to find optimal solutions to real problems.

Fast forward to 2017, and my work is focused almost exclusively on AI. I work with enterprises to help create their AI strategy—identifying opportunities for AI, finding the right solution or vendor and creating the roadmap for implementation. I don't do this as a technologist, but as someone that understands the capabilities of AI and how those capabilities might address business challenges and opportunities. There are plenty of people much cleverer than me that can create the algorithms and design the actual solutions, but those same people rarely understand the commerciality of business. I see myself as a 'translator' between the technologists and the business. And with AI, the challenge of translating the technology is orders of magnitude greater than with traditional IT. Which is why I wanted to write this book—to bring that understanding to where it can be used the best: on the front line of business.

So, this is not a book about the theoretical impact of AI and robots in 10 or 20 years' time and it is certainly not a book about how to develop AI algorithms. This is a book for practitioners of AI, people who want to use AI to make their businesses more competitive, more innovative and more future-proof. That will happen only if the business leaders and executives understand what the capabilities of the technology are, and how it can be applied in a practical way. That is the mission of this book: be as informed as possible about AI, but without getting dragged down by the technology, so that you can make the best decisions for your business. And it is also a heartfelt appeal: whatever you read or hear about AI, don't believe the hype.

2

Why Now?

A Brief History of AI

For anyone approaching AI now, the technology would seem like a relatively new development, coming off the back of the internet and 'big data'. But the history of AI goes back over 50 years, and includes periods of stagnation (often referred to as 'AI Winters') as well as acceleration. It is worth providing a short history of AI so that the developments of today, which really is AI's big moment in the sun, can be put into context.

I've already mentioned in the previous chapter the two people that are considered some of the key founders of AI, Marvin Minsky and Douglas Engelbert, who both originally worked at the Massachusetts Institute of Technology in Boston, USA. But the person who coined the phrase 'artificial intelligence' was John McCarthy, a professor at Stanford University in California. McCarthy created the Stanford Artificial Intelligence Lab (SAIL), which became a key focus area for AI on the west coast of America. The technology driving AI at that time would seem rudimentary when compared to the neural networks of today, and certainly wouldn't pass as AI to anyone with a basic understanding of the technology, but it did satisfy our earlier definition of "the theory and development of computer systems able to perform tasks normally requiring human intelligence", at least to a very basic level.

Much of the work in the early development of AI was based around 'expert systems'. Not wanting to disparage these approaches at all (they are still being used today, many under the guise of AI), they were really no more than 'if this then that' workflows. The programmer would lay out the knowledge of the area that was being modelled in a series of branches and

© The Author(s) 2018
A. Burgess, *The Executive Guide to Artificial Intelligence*,
https://doi.org/10.1007/978-3-319-63820-1_2

loops, with each branch depending on an input from the user or a rule. So, in a system designed to model a bank account recommendation process, the user would be asked a series of questions (employment status, earnings, savings, etc.) with each answer sending the process down different branches until it came to a conclusion. And, because that was essentially performing a task 'normally requiring human intelligence' it was, at the time, considered AI. Today it wouldn't really pass muster because it doesn't have any self-learning capability, which is a key facet of how AI is perceived to be defined.

Interestingly, even now, this same approach is being used in many of the chatbots that have proliferated across the internet. Most of these claim to use AI, and some do, but most are passive decision trees. There are a number of online chatbot platforms (most are free to use) that allow you to create your own bots using this approach and they do a reasonable job for simple processes (I actually built my own a while back—it took about half a day and was very basic—but it proved to me that it could be done by a non-technical person and that there was very little, if any, AI actually involved).

There were a couple of 'AI Winters', when advances in AI stagnated for a good number of years. Both of these were the result of over-inflated expectations and the withdrawal of funding. The first occurred between 1974 and 1980, triggered by three events: a 1973 report by Sir James Lighthill for the UK Government which criticised the 'grandiose objectives' of the AI community and its failure to get anywhere near reaching those objectives; and in the United States where the Mansfield Amendment required the Advanced Research Projects Agency (ARPA, now known as DARPA) to only fund projects with clear missions and objectives, particularly relating to defence, which AI at the time couldn't satisfy; and the perceived failure of a key project for ARPA that would have allowed fighter pilots to talk to their planes. These events meant that much of the funding was withdrawn and AI became unfashionable.

The second AI Winter lasted from 1987 to 1993 and was chiefly due to the failure of 'expert systems' to meet their over-inflated expectations in 1985, when billions of dollars were being spent by corporations on the technology. As with my own chatbot experience that I described above, expert systems ultimately proved difficult to build and run. This made them overly expensive and they quickly fell out of favour in the early 1990s, precipitated by the simultaneous collapse of the associated hardware market (called Lisp machines). In Japan, a 1981 program costing $850m to develop a 'fifth generation computer' that would be able to 'carry on conversations, translate languages, interpret pictures, and reason like human beings' failed to meet any of its objectives by 1991 (some remain unmet in 2017). And, although DARPA had started to fund AI projects in the United States in 1983 as a response to the Japanese efforts, this was withdrawn again in 1987 when new leadership at the Information Processing Technology Office, which was directing the efforts and funds of the AI, supercomputing and microelectronics projects, concluded that AI was 'not the next wave'. It dismissed expert systems as simply 'clever programming' (which, with hindsight, they were pretty close on).

I talk about these AI Winters in a little detail because there is the obvious question of whether the current boom in AI is just another case of over-inflated expectations that will lead to a third spell of the technology being left out in the cold. As we have seen in the previous chapter, the marketing machines and industry analysts are in a complete froth about AI and what it will be capable of. Expectations are extremely high and businesses are in danger of being catastrophically let down if they start to believe everything that is being said and written. So, we need to understand what is driving this current wave and why things might be different this time.

From a technology point of view, the only two words you really need to know for now are 'machine learning'. This is the twenty-first century's version of expert systems—the core approach that is driving all the developments and applications (and funding). But, before I describe what machine learning is (in non-technical language, of course), we need to understand what all the other forces are that are contributing to this perfect storm, and why this time things could be different for AI. There are, in my mind, four key drivers.

The Role of Big Data

The first driver for the explosion of interest and activity in AI is the sheer volume of data that is now available. The numbers vary, but it is generally agreed that the amount of data generated across the globe is doubling in size every two years—that means that by 2020 there will be 44 zettabytes (or 44 trillion gigabytes) of data created or copied every year. This is important to us because the majority of AI feeds off data—without data this AI would be worthless, just like a power station without the fuel to run it.

To train an AI system (like a neural network) with any degree of accuracy you would generally need millions of examples—the more complex the model, the more examples are needed. This is why the big internet and social media companies like Google and Facebook are so active in the AI space—they simply have lots of data to work with. Every search that you make with Google—there are around 3.5 billion searches made per day—and every time you post or like something on Facebook—every day 421 billion statuses are updated, 350 million photos are uploaded and nearly 6 trillion Likes are made—more AI fuel is generated. Facebook alone generates 4 million gigabytes of data every 24 hours.

This vast volume of data is then consumed by the AI to create value. To pick up again on the simple example I used in the previous chapter, to train a DNN (essentially a machine learning AI) to recognise pictures of dogs you would need to have many sample images of dogs, all labelled as 'Dog', as well as lots of other images that didn't contain dogs, all labelled 'No Dog'. Once the system had been trained to recognise dogs using this set of data (it could also go through a validation stage where the algorithm is tuned using a sub-set of the training data), then it will need to be tested on a 'clean', that is, unlabelled, set of images. There are no strict guidelines for how much testing data is required but, as a rule of thumb, this could represent around 30% of the total data set.

These huge amounts of data that we create are being exploited every minute of the day, most of the time without our knowledge (but implicit acceptance). Take, for example, your Google searches. Occasionally, as you type in your search term you may spell a word incorrectly. Google usually offers you results based on a corrected or more common spelling of that word (so for example if I search for 'Andrew Durgess' it shows me results for Andrew Burgess), or you can choose to actually search for the uncommonly spelled version. What this means is that Google is constantly collecting data on commonly misspelt versions of words and, most importantly, which corrections that it suggests are acceptable or not. All the data is then used to continually tune their AI-powered spell checker. If, in my example, there was actually someone called Andrew Durgess who suddenly became famous tomorrow so that many people were searching for his name, then the correction that Google used on my name would be quickly phased out as less and less people accepted it and instead clicked on 'Search instead for Andrew Durgess'.

But it's not just social media and search engines that have seen exponential increases in data. As more and more of our commercial activities are done online, or processed through enterprise systems, more data on those activities will be created. In the retail sector our purchases don't have to be made online to create data. Where every purchase, whilst not necessarily connecting it to an identified buyer, is recorded, the retailers can use all that data to predict trends and patterns that will help them optimise their supply chain. And when those purchases can be connected to an individual customer, through a loyalty card or an online account for example, then the data just gets richer and more valuable. Now the retailer is able to predict what products or services you might also want to buy from them, and proactively market those to you. If you are shopping online, it is not only the purchase data that is recorded—every page that you visit, how long you spend on each one and what products you view are all tracked, adding to the volume, and the value, of data that can be exploited by AI.

After the purchase has been made, businesses will continue to create, harvest and extract value from your data. Every time you interact with them through their website or contact centre, or provide feedback through a third-party recommender site or social media, more data is created that is useful to them. Even just using their product or service, if it is connected online, will create data. As an example, telecoms companies will use data from your usage and interactions to try and predict, using AI, whether you will leave them soon for a competitor. Their 'training data' comes from the customers that have actually cancelled their contracts—the AI has used it to identify all of the different characteristics that make up a 'churn customer' which it can then apply to the usage and behaviours of all of the other customers. In a similar way, banks can identify fraudulent transactions in your credit card account simply because they have so much data available of genuine and not-so-genuine transactions (there are roughly 300 million credit card transactions made every day).

Other sources of 'big data' are all the text-based documents that are being created (newspapers, books, technical papers, blog posts, emails, etc.), genome data, biomedical data (X-rays, PET, MRI, ultrasound, etc.) and climate data (temperature, salinity, pressure, wind, oxygen content, etc.).

And where data doesn't exist, it is being created deliberately. For the most common, or hottest, areas of AI complete training sets of data have been developed. For example, in order to be able to recognise handwritten numbers, a database of 60,000 samples of handwritten digits, as well as 10,000 test samples, has been created by the National Institute of Standards (the database is called MNIST). There are similar databases for face recognition, aerial images, news articles, speech recognition, motion-tracking, biological data and so on. These lists actually become a bellwether for where the most valuable applications for machine learning are right now.

The other interesting aspect about the explosion and exploitation of data is that it is turning business models on their heads. Although Google and Facebook didn't originally set out to be data and AI companies, that is what they have ended up as. But what that means now is that companies are being created to harvest data whilst using a (usually free) different service as the means to do it. An example of this, and one which uses data for a very good cause, is Sea Hero Quest. At first sight this looks like a mobile phone game, but what it actually does is use the data from how people play the game to better understand dementia, and specifically how spatial navigation varies between ages, genders and countries. At the time of writing 2.7 million people have played the game which means it has become the largest dementia study in history. Commercial businesses will use the same approach of a 'window' product or service that only really exists to gather valuable data that can be exploited elsewhere.

The Role of Cheap Storage

All the data that is being created needs to be stored somewhere. Which brings me on to the second driver in favour of AI: the rapidly diminishing cost of storage, coupled with the speed that data can be accessed and the size of the machines that store it all.

In 1980 one gigabyte of storage would cost on average $437,500. Five years later that had dropped to around a quarter, and by 1990 it was a fortieth of the 1980 price, at $11,200. But that was nothing compared to the subsequent reductions. At the turn of the century it was $11.00, in 2005 it was $1.24 and in 2010 it was 9 cents. In 2016, the cost stood at just under 2 cents per gigabyte ($0.019).

Because of all that data that is generated through Facebook that I described above, their data warehouse has 300 petabytes (300 million gigabytes) of data (the amount of data actually stored is compressed down from that which is originally generated). Accurate numbers are tricky to come by, but Amazon's Web Services (its commercial cloud offering) probably has more storage capacity than Facebook. It is this sort of scale that results in a sub–2 cent gigabyte price.

It's not just costs that have shrunk; size has too. There is a photograph from 1956 that I use in some of the talks I give of an IBM hard drive being loaded onto a plane with a fork-lift truck. The hard drive is the size of a large shed, and only has a capacity of 5 Mb. Today that would just about be enough to store one MP3 song. Amazon now has a fleet of trucks that are essentially huge mobile hard drives, each capable of storing 100 petabytes (the whole of the internet is around 18.5 petabytes). At the time of writing IBM have just announced that they have been able to store information on a single atom. If this approach can be industrialised it would mean that the entire iTunes library of 35 million songs could be stored on a device the size of a credit card.

The Role of Faster Processors

All this cheap storage for our massive data sets is great news for AI, which, as I have hopefully made very clear, lives off large amounts of data. But we also have to be able to process that data too. So, the third driver in AI's perfect storm is faster processor speeds.

This is where the often-cited Moore's Law comes in. The founder of Intel, Gordon Moore, predicted in 1965 that the number of transistors that could fit onto a transistor would double every year. In 1975, he revised this to a doubling every two years. It was actually David House, an Intel executive, who came up with the most commonly used version which is that chip performance (as an outcome of more and faster transistors) would double every 18 months. Although there have been deviations from this trend, particularly in the past few years, it means that the processors being used today are orders of magnitude faster than during the last AI Winter.

One of the curious things to come out of technology for AI is that traditional computer chips (central processing units, or CPUs) are not ideally suited to the crunching of large data sets, whereas GPUs (graphical processing units), which had been developed to run demanding computer visualisations (such as for computer games), were perfect for the job. Therefore NVidia, a maker of GPUs, has taken most of the market for computer chips in the world of AI.

So, faster AI-friendly processors mean that more complex problems, that use more data, can be solved. This is important because managing and processing all that data does take time—the systems are fine at the second part of the process, that of scoring and making a decision based on the learnt data, but the training part can be a slog. Even relatively simple training sessions need to run overnight, and more complex ones can take days. So any improvement in processor speed will help enormously in the usefulness of AI systems both in the original development and design of the models but also on a day-to-day basis so that the systems can use the most up-to-date models. Being able to provide real-time training as well as real-time decision making is one of the last frontiers of really useful AI.

The Role of Ubiquitous Connectivity

The final driver working in AI's favour is connectivity. Clearly the internet has had a huge enabling effect on the exploitation of data, but it is only in the past few years that the networks (both broadband and 4G) have become fast enough to allow large amounts of data to be distributed between servers and devices. For AI this means that the bulk of the intensive real-time processing of data can be carried out on servers in data centres, with the user devices merely acting as a front end. Both Apple's Siri (on the iPhone) and Amazon's Alexa (on the Echo) are prime examples of very sophisticated AI applications that exploit processing power in data centres to carry out the bulk of the hard work. This means there is less reliance on the device's processor but it does put a burden on having the network available and effective enough.

It's not only the real-time processing where the internet can provide benefits. The sheer effort in training an AI model, with each training run taking days or weeks on 'standard' hardware, can be significantly sped up using a cloud-based solution that has specialised hardware available.

Better communication networks can also help AI systems in other ways. The data sets that I mentioned in the previous section are huge but are made available, usually publicly, to many users to help train their systems—that wouldn't have been nearly as easy previously.

Also, AI systems can be connected to each other using the internet so that they can share learnings. A program run by a consortium of Stanford University, University of California at Berkeley, Brown University and Cornell University, called Robobrain, uses publicly available data (text, sounds, images, film) to train AI systems which can be accessed by other AI systems. And, of course, like any good AI system, the 'recipient' systems will feed back everything that they learn back into Robobrain. The challenge for Robobrain, which mirrors the overall challenge of AI, is that systems tend to be very narrow in their focus whereas Robobrain wants to be all things to all robots (or 'multi-modal' to use the lingo).

About Cloud AI

Nowhere has all four of these drivers come together more effectively than in the concept of Cloud AI. The idea of AI-as-a-service, where the heavy-lifting of the AI is done on the cloud and on-demand, is now a key catalyst for the democratisation of AI. Many of the large technology firms, such as Google, IBM and Amazon, all have cloud solutions for AI that offer easily accessible APIs (Application Programming Interfaces, which are essentially standardised access points to programming capabilities) for developers to create AI 'front-ends' out of them. IBM's Watson—its heavily branded and lauded AI capability—is 'just' a series of APIs, with each one carrying out a specific function, such as speech recognition or Q&A. Google's TensorFlow is an open-source AI platform that offers similar capabilities, as well as additional features such as pre-trained models.

What this means for business, and specifically any entrepreneur wanting to start an AI business, is that the value of the AI is not going to be where everyone assumes it be: in the algorithm. If every new customer-service focused AI business used the open-source speech recognition algorithm from, say, Amazon, then the competitive advantage would have to lie with the quality of the training data, the way that the algorithm is trained or how easily the resulting system is to use. When you connect to an IBM Watson API, for example, there is still a lot of work to do training it before you can derive any real value.

Clearly some AI businesses will create competitive advantage from their algorithm, but they must compete with the other firms who use an off-the-shelf one. It's fairly straightforward, as I have done myself, to download a free, open-source Named Entity Recognition algorithm (which is used to extract names such as people, places, dates, etc., from a body of text) from a university website (in my case it was Stanford's), feed it some sample text and get it to return a reasonable answer. But that is not yet a viable AI solution, let alone a business. To make it commercial I need to train and tune the algorithm using as much data as I can get my hands on and create a user-friendly interface for it, and that is where the real skills—in data science, in optimisation, in user experience—come into play. Put all that together and you may have the seed of an AI business.

For the executive looking to use AI in their business (and the fact that you are reading this book would suggest that is the case) the provenance of the value in an AI company is important to understand. Why should you choose one vendor over another, if they say they can do exactly the same thing? Getting to the heart of the differences—is it down to the algorithm, the data, the ease of implementation, the ease of training, the ease of use and so on?—will allow you to make the right choice for your business. At the moment, there are far too many smoke-and-mirrors AI businesses that are based purely on open-source algorithms and built by very clever but inexperienced 20-somethings. They will be riding the hype of AI but will offer little in the way of long-term value. My point being that you can create a successful AI business based on great open-source algorithms, but you need all the other aspects to be great too. As I said right at the start of this book: don't believe the hype.

I'll be covering these selection considerations, as well as the buy-versus-build question, in more detail later in the book.

What Is Machine Learning?

These four developments then—big data, cheap storage, faster processors and ubiquitous networks—are what is driving the acceleration and adoption of AI in business today. If you were to take any one of those away, AI would struggle to perform, and we'd probably still be in an AI Winter. Each one has also been helping drive development in the others—if we didn't have such cheap storage then we couldn't keep all that data which means there would be no need for faster processors, for example. But, at the centre of all this activity, feeding off each of the drivers, is machine learning, one of the most popular AI approaches.

It's important to have an appreciation of what machine learning is and what it does, even if you don't need to understand how it actually works. As the name suggests, it is the machine that does all of the hard computational work in learning how to solve a problem—the machine, rather than a human being, essentially writes the 'code'. The human developer plants the seed by defining the algorithm or algorithms that are to be used, and then the machine uses the data to create a solution specific to it. We have already seen that the problem could be undefined initially (as in unsupervised learning, where patterns or clusters are found in data) or defined (and therefore trained with large data sets to answer a particular question, which we know as supervised learning). For machine learning it is usually easier to think about a supervised learning example, such as my go-to case of being able to tell the difference between pictures of dogs and pictures of cats.

In the previous chapter, I introduced the idea of DNNs, which is the 'architecture' that machine learning uses. A DNN will consist of a number of different layers; the more complex the problem being solved, the more layers are needed. (More layers also mean that the model is much more complicated, will require more computing power and will take much longer to resolve itself.) The Input Layer takes the data in and starts to encode it. The Output Layer is where the answer is presented—it will have as many nodes as there are classes (types) of answer. So, in my dog versus cat example, there would be two output nodes, one for dog and one for cat (although we could define three if we had pictures that had neither a dog nor a cat in) (Fig. 2.1).

In between the Input Layer and the Output Layer are the Hidden Layers. It is here that all the hard work happens. Each Hidden Layer will be looking for different features within the data with increasing complexity, so with image recognition different layers will look for outlines, shadows, shapes, colours and so on. In these Hidden Layers, there will be more nodes ('neurons') than the Input or Output Layers, and these nodes will be connected to each other across each layer. Each of these connections will have a certain 'weight' or strength which determines how much of the information in one node is taken into the next layer: a strong link, which had, through training, led to a 'right' answer at the Output Layer, will mean that information is propagated down through to the next layer. A weak link with a low weighting, which had led to a 'wrong' answer in training, will not pass as much information down (Fig. 2.2).

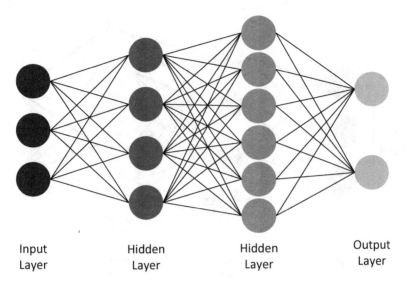

Fig. 2.1 Basic neural network

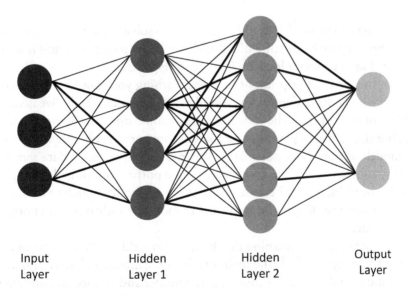

Fig. 2.2 Training a neural network

As the model is being trained with more and more data, the weights are continually adjusted (this is the machine learning) until it reaches an optimal solution. The more data we have fed into the model, the more chances the machine has to refine the weightings (but the harder it has to work), and therefore the more accurate the solution will be. The 'matching function' (the

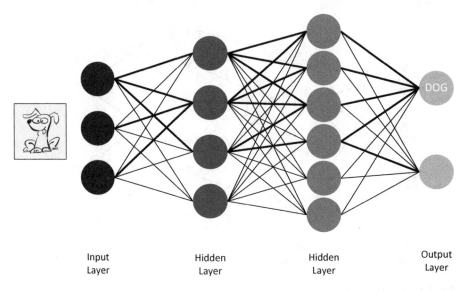

Input   Hidden   Hidden   Output
Layer   Layer   Layer   Layer

Fig. 2.3 A trained neural network

final version of the model) can then be used to solve for a new piece of data; for example, I give it a picture of a dog that it hasn't seen before and it should then be able to correctly identify it as a dog (Fig. 2.3).

From this brief description of machine learning you should be able to see that it is wholly dependent on the four drivers I have described—we need lots and lots of data to train the hidden layers to come up with the right weightings, but that means we need to be able to store our data cheaply and process our models as quickly as possible, as well as needing access to data sets from as many sources as possible. Without any one of these, machine learning just wouldn't be viable, either because it wouldn't be accurate enough or it wouldn't be easy enough to design and implement (and it's already difficult enough in the first place).

It's not just machine learning that benefits from this 'perfect storm' of technology. Other types of AI, the main one being 'symbolic AI', have also accelerated and found new life through faster, cheaper and connected computers.

The Barriers to AI

So, with all this data, cheap storage, whizzy processors and inter-connectivity, we are certain to never see another AI Winter, right? Well, there are a few things that could spoil the AI party.

The biggest barrier to AI achieving escape velocity in my opinion is the over-inflation of expectations. Far too much is being written and said about what AI might be capable of, especially when Artificial General Intelligence is considered and not the Narrow AI that we have today. Both the previous AI Winters have been precipitated by excessive and unrealistic expectations, and therefore they need to be managed carefully. A key motivation for me to write this book was to 'bring AI down to earth' so that it could be assessed for its real value today and tomorrow, and not in 10 or 20 years' time.

Another factor that will affect the uptake of AI, which has been fuelled by the hype, is a general fear of what changes AI might bring, especially when it could mean fundamentally changing the way that people work. There have been many other seismic shifts in ways of working, including the introduction of personal computers and outsourcing, but AI has the potential to do this on a much greater scale: speculative reports by academics and institutions repeatedly report the decimation of 'white collar' jobs and the 'hollowing out' of the middle classes. All of this may or may to be true, but the fear that it generates leads to a natural defensiveness from those in businesses that may be affected by the implementation of AI.

A third aspect, which is connected to the first two, is ignorance. How can people expect to get real value from implementing AI if they don't have a good enough understanding of what it is? The hype doesn't help, but AI is inherently a complex subject, and it's not an easy task for a non-technical business person to understand it. The maxim that a little knowledge is a dangerous thing is certainly true for AI, but also having all the knowledge is not useful unless you want to be a data scientist or AI developer. There is a 'Goldilocks' level of understanding AI that this book aims to provide—it should be just enough that you can make the most out of AI but not too much that you become overwhelmed with technical detail and jargon.

The final thing that could put the brakes on AI is regulation. As I've mentioned already, much of the computational work that is done by an AI system is hidden from the developer or user—there is (generally) no audit trail that details how it came to a certain decision. If I fed a complex credit decisioning AI system with lots of loan applications as training data and told it which had been approved and which hadn't, it could then make a recommendation of whether to approve or reject a new loan. But no one would really know why, and that causes a problem for regulators who need to see that decision process written down. There are some approaches which can mitigate this but the opaqueness of AI can be a real challenge to its usefulness.

I'll cover these aspects in more detail in Chap. 8, but I think it's important that you have an early appreciation of them at this point.

Some AI Case Studies

Although it may be fuelled by perhaps too much hype, there is certainly momentum in the AI market right now. But who is benefiting from all these advancements? Is it just research labs and some start-ups that have had success playing computer games, or is there something more? Is AI providing value to businesses today?

In my work, I see plenty of examples of AI being used in businesses to provide insights and efficiencies, adding value that would be beyond the means of humans to create.

One of the sweet spots for AI at the moment is in customer services. At a train operating company in the United Kingdom they use AI to categorise emails that come in from their customers. The challenge, of course, is for the AI to be able to 'read' the email, which has been written freeform, and determine what the customer actually wants. For example, if I couldn't get a seat on the 08:06 from Euston Station in London to Glasgow Central, for example, I would email the train company, saying something like "I can't believe it – I paid all that money for a ticket on the 08:06 from London to Glasgow and I couldn't even get a seat. You've really excelled yourselves this time!" All the salient information is extracted from my unstructured text and validated (Was there a train at 08:06? Have other people complained about the same service? Is the customer a regular complainer? etc.) so that the AI can immediately route my query to the relevant person in the customer service organisation. Of course, my wording in the second sentence is sarcastic in tone, so the AI needs to understand the difference between my complaint and an actual compliment. What it all means is that when the human agent logs in, they are the right person to deal with my query and they have everything at hand to respond.

In a similar vein, a UK-based insurance firm that provides a claims processing service to their own insurance customers has automated the input of unstructured and semi-structured data (incoming claims, correspondence, complaints, underwriters' reports, cheques and all other documents relating to insurance claims) so that it is directed into the right systems and queues. Using an AI solution, a team of four people process around 3,000 claims documents per day, 25% of which are paper. The AI automation tool processes the scanned and electronic documents automatically, identifies claim information and other metadata, and deposits the results in databases and document stores ready for processing by the claims handlers and systems. It also adds service metadata so performance of the process can be measured end to end. Some documents can be processed without any human intervention, and others need a glance from the human team to validate the AI's decisions or fill in missing details.

A good example of AI in the legal sector is ROSS, a system built around IBM's Watson. ROSS actually uses a number of AI capabilities including NLU, Search and Optimisation. So, when a lawyer has a matter that requires some specific research (e.g. what is the precedent when an employee is fired for gross misconduct within a few days of the end of their probationary period?), she can turn to ROSS and type in that query in plain English. ROSS will then interrogate the whole corpus of employment law and provide answers ranked by relevancy back to the lawyer within seconds. The alternative would be hours of research by the lawyer, a junior lawyer and/or a paralegal, and likely without being able to examine every relevant document available. Like all good AI systems, ROSS is self-learning—the lawyer can evaluate the quality of the responses that ROSS returns and therefore allow it to provide better answers in the future.

Some companies have used AI to create whole new lines of business. Charles Schwab, the US bank and brokerage firm, has created an investment vehicle, called Schwab Intelligent Portfolios, which uses AI to manage its clients' portfolios (these are sometimes called 'robo-advisors'). It focuses on low-cost exchange-traded funds and has no advisory fees, account service fees or commission fees. Since it launched in 2015, a few other firms have appeared with a similar model, including Betterment, Wealthfront and FutureAdvisor (although these charge a very small admin fee). The attractiveness of the low or zero fees and the simplicity of dealing with an AI rather than a person have meant that these sorts of services have proved popular amongst beginner investors, thus providing new clients for the banks that they normally wouldn't have been able to access.

Conclusion

So, AI is being used successfully by businesses around the world. The perfect storm of big data, cheap storage, faster processors and ubiquitous connectivity has allowed researchers to exploit the near-magic that is machine learning. But AI has managed to escape the lab and is being commercialised by both start-ups and internet giants. Real opportunities exist for executives to use AI's capabilities to deliver new sources of value into their business as well as to challenge and disrupt their existing business models. In order to do that they need to understand what those capabilities are and how they can be used.

The Academic's View

This is an extract from an interview with Dr Will Venters, Assistant Professor in Information Systems at the London School of Economics.

AB What's the basis of all the fuss around AI right now?

WV There are a few trajectories that are particularly important. The underlying algorithms have become much smarter and are used more effectively, but that's not actually that significant. What has really made the difference is the processing capacity of the computer chips as well as the ability to handle much bigger volumes of data at scale.

Previously we saw IBM's Big Blue beat Kasparov at chess, then Watson won at Jeopardy!, but essentially these are just systems that can search massive databases very quickly. But what AlphaGo is doing (beating the world's number one player in the Chinese game of Go) is amazing—it watched, then played millions of games of Go against itself. The complexity and the sheer quantity of data that it was able to analyse is phenomenal—I'm sure the future of AI will be around this sort of capability.

It's interesting if you look at data complexity versus task complexity for processes that computers want to take on: we are now able to manage situations that are complex in both of these aspects. Managing elevators is a complex task, but the data is relatively simple. Self-driving cars are not doing complex things (just driving and turning and stopping) but the data they need to do that is vast. But, we are now able to manage situations that are complex in both data and task aspects. Suddenly we have the mechanisms to do it.

AB You also mentioned processing power as being a key factor?

WV Yes, GPUs [Graphical Processing Units] have made a significant contribution to this. But also the widespread use of Cloud Computing, which allows businesses to manage big data without the associated infrastructure investments and risks.

Lots of companies have invested in data infrastructure, either their own or through the cloud, and AI allows them to utilise this to the full.

AB Do you think the hype helps or hinders AI?

WV AI is certainly hyped, but it is also productive. It leads to new innovations and forces businesses to discuss big questions. Hype shouldn't only be seen in the negative.

AB What do you see as the most valuable use cases for AI?

WV AI can offer the opportunity to deal with 'messy' data – data that is fuzzier, such as photographs, natural language speech and unstructured documents.

Big Data promised to deliver all of this but ultimately it didn't because it required armies of statisticians. Robotic Process Automation is great for dealing with processes that are clean and structured – as soon as it becomes messier then AI can step in and overcome that.

AB   What do you think businesses need to do if they are to exploit AI to the full?

WV   They first need to look at the business case, and not the technology. But they also need to consider the governance and to understand how they manage the inherent risks. Humans have an ethical compass which, of course, AI doesn't. It will need some level of oversight because of the lack of transparency with the AI models. They must also make sure the system is not biased – it's very easy to bake the bias into the company's business. And these risks scale very quickly because the underlying data is so much bigger.

AB   What do you see for the future for AI?

WV   I think we will see even better algorithms to deal with the bigger volumes of data. But, of course, AI is a victim of its own success – there is an inherent disillusionment as each exciting step forward then becomes common place. You just need to look at advances like [Apple's] Siri and [Amazon's] Alexa. The term AI will eventually become redundant and people will cease to be questioning about it.

But until that day companies will invest in AI wisely as well as inappropriately. There is so much value to be gained that companies should think deeply about their approach, take expert advice and make sure they are not left behind.

3

AI Capabilities Framework

Introduction

It's impossible to get value out of something if it is not understood, unless it's by some happy accident. In the world of AI there are no happy accidents—everything is designed with meticulous detail and with specific goals in mind. Therefore, the only way to truly make the most of AI in business is by having a reasonable understanding of it. The challenge, of course, is that is fiendishly complicated and full of complex mathematics, and is certainly not within the bounds of what an 'ordinary' business person would be expected to grasp.

My approach in this book is to understand AI from the perspective of what it can do, that is, what are its capabilities in the context of real world problems and opportunities? In order to do this, I have developed a framework that 'boils down' all of the complexity into eight capabilities. In theory, any AI application should fit into one of these, which allows the AI layman to quickly assess and apply that application into something that relates to their business. Conversely, if you have a specific business challenge, the AI framework can be used to identify the most appropriate AI capability (or capabilities) that could address that need. Having said that, I'm sure any AI scientist or researcher would be able to pick some holes in the framework. It's certainly not bomb-proof: it is intended to be a useful tool for business executives to help them get the most value possible out of AI.

There are many ways to look at AI, some of which I have already discussed in earlier chapters: Supervised versus Unsupervised Learning; Machine Learning versus Symbolic Learning; Structured Data versus Unstructured Data;

© The Author(s) 2018
A. Burgess, *The Executive Guide to Artificial Intelligence*,
https://doi.org/10.1007/978-3-319-63820-1_3

Augmentation versus Replacement. All of these are valid perspectives, and all can be understood to a reasonable enough degree within the capability framework.

So, with each of the capabilities I will explain whether it will require supervised or unsupervised learning (or both), the type of AI that is usually used to deliver the capability (e.g. machine learning), whether it processes structured or unstructured data and, most importantly, how it benefits businesses. In subsequent chapters I will draw out use cases for each, grouped into the key themes of enhancing customer service, business process optimisation and enhanced decision making.

Another perspective for AI, which I introduced in Chap. 1, is really the meta-view, and probably the most important to understand first: these are the three basic things that AI tries to achieve: *capturing information*; determining *what is happening* and understanding *why it is happening*. Each of my eight capabilities fits into one of these 'AI Objectives' (Fig. 3.1):

Capturing information is something that our brain does very well but machines have historically struggled with. For example, the ability to recognise a face is something that has evolved in humans since our earliest existence—it is a skill that allows us to avoid danger and create relationships, and therefore a great deal of brain power and capacity is used to work on this problem. For machines, the process is very complex, requiring huge volumes of training data and fast computer processors, but today we have this capability on our desktop computers and even mobile phones. It may not be very fast or that accurate yet but AI has been fundamental in achieving this.

Most of the examples of capturing information are where unstructured data (such as a picture of a face) is turned into structured data (the person's name). Capturing information is also relevant to structured data, and AI comes into its own when that data is big. Again, the human brain is very good at identifying patterns in data (the impact of a manager on the success of a football team, for example) but when there are hundreds of variables and millions of data points we lose sight of the forest and can only see the trees. AI is able to find patterns, or clusters, of data that would be invisible to a human. These clusters have the ability to provide insights within the data that have real value to a business. For example, AI can find patterns between purchases and customer demographics that a human would either take years to unearth, or never have thought of in the first place. (This associated idea of the AI

Fig. 3.1 AI objectives

being 'naive' to the data is something that I will return to later, but is an important concept to bear in mind.)

The next AI Purpose is where it tries to determine **what is happening**, and is usually a consequence of where information has already been captured by AI. For example, we could have used Speech Recognition to extract (from a sound file or a live conversation) the words that someone was speaking, but at that point we would only know what all the individual words were rather than what the person was actually trying to say. This is where Natural Language Understanding (NLU) comes in—it takes the words and tries to determine the meaning or intent of the complete sentences. So, we have gone from a digital stream of sound to a set of words (e.g. 'I', 'want', 'to', 'cancel', 'my', 'direct', 'debit', 'for', 'mortgage', 'protection') to working out that this person 'wants to cancel her Direct Debit for her mortgage protection insurance'.

We can then use other capabilities in this category to take that request further. For example, we could use an Optimisation approach to help the customer understand that if they were to cancel that Direct Debit they would also probably need to cancel the insurance policy that it is associated with. And then we could use a Prediction capability to work out that this customer may be about to leave the bank and go to a competitor (because the AI has determined from lots of other similar interactions that cancelling Direct Debits is an indicator of potential customer churn).

So, from a simple interaction with a customer we are able to use different AI 'capturing information' and 'what is happening' capabilities to build up a useful picture as well as satisfying the customer's request. But, in the example above, although the AI has been able to identify this customer as a 'churn risk' it really doesn't **understand** what that means. All it has done is correlate one set of data (customer requests) with another (customers who leave) and then apply that to a new data point (our customer's Direct Debit cancellation request). To the AI system the data might as well be about ice cream flavours and the weather. It has no idea of the *concept* of Direct Debits just as much as it has no idea of the concepts of banks or customers. The AI capabilities that we have today and will have in the near future (and maybe even *ever*) do not have the ability to *understand*. It is really important to be able to differentiate between the very *narrow* things that different AIs can do (usually better than humans) and the *general* intelligence that comes with understanding and relating different concepts—something our brain is brilliant at doing.

Now that we have the three AI purposes defined, and we understand that only the first two are relevant to us today, we can start to look in more detail at each of the eight AI capabilities in the Framework.

Image Recognition

One of the most active areas of research in AI at the moment is that of Image Recognition. This is a prime example of where the four key drivers that I described in Chap. 2 have aligned to catalyse the technology. Image Recognition is based on Machine Learning and requires thousands or millions of sample images for training; therefore, it also needs lots of storage for all the data, and fast computers to process it all. Connectivity is also important to enable access to the data sets, many of which are publicly available (in Chap. 2 I mentioned image data sets for handwritten numbers and faces, but there are many more including aerial images, cityscapes and animals).

Images, of course, fall under the category of unstructured data. So, what sort of applications would you use Image Recognition for? There are three main types:

Probably the most popular application is identifying what is in the image, sometimes referred to as *tagging*. I've used this example a few times already—find out if the picture contains a dog or a cat. It is often used to moderate photographs by trying to identify pornographic or abusive images, but can also be used to group photos with similar tags ('these are all pictures taken at the beach'). This photo tagging is a prime example of supervised learning—the AI is trained on thousands, or millions, of tagged photographs—which is why companies that have access to large volumes of images, such as Google and Facebook, have some of the most advanced systems in this area.

Another use of Image Recognition is to find images that are *similar to other images*. Google's Reverse Image Search is a popular use of this approach; you simply upload a picture and it will search for pictures that look similar to your original (this technique is often used to identify the use of news photos that have deliberately been used out of context). Unlike photo tagging, this is mainly an example of unsupervised learning; the AI doesn't need to know what is in the picture, just that it looks like another picture. (A simple way to think about this is to understand that the AI converts the image file into a long series of numbers—it then looks for other images that have a similar series of numbers.)

The final application for Image Recognition is to find *differences between images*. The most common, and beneficial, use of this is in medical imaging. AI systems are used to look at scans of parts of the body and identify any anomalies, such as cancerous cells. IBM's Watson has been a pioneer in this space and is used to support radiologists in their work. The approach uses supervised learning to tag, for example, X-rays as healthy or unhealthy. Based on the algorithmic model that is built up, new images can be evaluated by the AI to determine whether there is a risk to the patient. Interestingly, it has been reported that Watson is more accurate at spotting melanomas than when it is done manually (Watson has 95% accuracy compared to an average of between 75% and 84% for humans).

Image Recognition, probably out of all the capabilities I describe in this chapter, is the most data hungry. Images are inherently unstructured and very variable and therefore require very large amounts of data to train them effectively. Pinterest, the website that allows users to create 'boards' of their favourite images, has used all of the data from their hundreds of millions of users to help develop their systems further (it does a good job of finding similar images to ones that you have posted, even if they are not tagged and are pretty abstract) but also to develop new AI applications such as Pinterest Lens which enables you to point your phone camera at an object and it will return pictures of objects that are visually similar, have related ideas or are the same object in other environments.

Other applications of Image Recognition are not so altruistic. A website in Russia, where privacy laws are more lax than most Western countries, enables its users to identify people in the street using their phone camera. Find Face, as the website is called, exploits the fact that all 410 million profile pictures on the country's most popular social media site, VK, are made public by default. Therefore, it is able to match, to an apparent accuracy of 70%, a face that you point your camera at with a VK profile. The intended, and rather bawdy, uses of this application are clear from the pictures of women on the website's homepage, but it does demonstrate how well image recognition technology can work if it has a large enough training set to work from.

Image recognition is still at a relatively immature stage—many useful things are possible now, but the potential is much greater. The use of still and moving images in our daily lives and in the business world is increasing exponentially and therefore the ability to index and extract meaningful data from these will be an increasingly important application.

Speech Recognition

Speech recognition, sometimes referred to as speech-to-text, is usually the first stage in a string of AI capabilities where the user is providing instructions by voice. It takes the sounds, whether live or recorded, and encodes them into text words and sentences. At this point other AI capabilities such as NLU would be needed to determine the meaning of the encoded sentences.

Speech recognition has benefited enormously from the development of DNNs although some of the more 'traditional' AI approaches (usually using something called a Hidden Markov Model or HMM) are still widely used today, mainly because they are much more efficient at modelling longer sections of speech. Just like for the images, the input data is unstructured, with the technology using supervised learning to match the encoded words with tagged training data (as such there are a number of publicly available speech-related training sets).

There are many challenges to an efficient and accurate speech recognition system, as most readers will have experienced themselves whilst trying to get their smartphone to understand a voice command from them. One of the main challenges is the quality of the input—this could be because of a noisy environment or because the voice is on the other end of a telephone line—accuracy on the phone drops from a best-in-class Word Error Rate (WER) of around 7% to more than 16%. (Human accuracy is estimated to be around 4% WER.)

Other challenges include obvious things like different languages and regional accents but a key consideration is the size of the vocabulary that will be required. For a very specific task, say checking your bank balance, the vocabulary will be very small, perhaps ten or so words. For systems that are expected to answer a wide range of questions, such as Amazon's Alexa, then the vocabulary will be much larger, and therefore present a greater challenge for the AI.

A wider vocabulary requirement also means that context becomes more challenging to determine. Context is important in speech recognition because it provides clues to what words are expected to be spoken. For example, if we hear the word 'environment' we would need to understand its context to determine whether the speaker meant 'our immediate surroundings' or 'the natural world'. DNNs, and a particular type of DNN called a Recurrent Neural Network, are very good at looking backwards and forwards through the sentence to constantly refine the probability that it is one meaning or another.

It is worth pointing out that speech recognition is a slightly different concept to voice recognition. Voice recognition is used to identify people from their voices and not to necessarily recognise the words they are saying. But many of the concepts used in the two applications are similar.

Speech recognition, and its cousin NLU, are seeing plenty of development activity as people become more used to the technology through their smartphones, and therefore more comfortable and confident with talking to machines. As a 'user interface', speech recognition will likely become the dominant input method for many automated processes.

Search

I use the word 'Search' in a specific sense here (another common term for this is 'Information Extraction'). What this AI capability is all about is the extraction of structured data from unstructured text. So, just as the Image Recognition and Speech Recognition capabilities worked their 'magic' on pictures and sounds, so the Search capability does with documents.

There are strong elements of Natural Language Processing (NLP) used in this capability to extract the words and sentences from the passages of text, but I prefer to use the term 'Search' because it better describes the overall capability. Later on in this chapter you will hear about Natural Language *Understanding*, which is generally described as a sub-set of NLP, but, in my mind, is a capability in itself which complements the outputs of Speech Recognition and Search.

AI Search, which is almost exclusively a supervised learning approach, works on both unstructured and semi-structured data. By unstructured data I mean something like a free-form email or a report. Semi-structured documents have some level of consistency between different instances but will have enough variability to make it extremely difficult for a logic-based system (such as a Robotic Process Automation [RPA] robot) to process.

A good example of a semi-structured document is an invoice. An invoice will generally have the same information as another invoice, such as the supplier's name, the date and the total value. But one may have a VAT amount on it, another may have a sub-total and a total, another may have the address written in the top right-hand corner rather than the top left, another may have the supplier's name written slightly different and another may have the date written in a different format.

The traditional way of getting the information off the invoice and into the appropriate 'system of record' would be to use an Optical Character Recognition (OCR) system and a template for each different version of the invoices (which could run into the hundreds) so that it knew where to find each piece of data. An AI system, once trained on a sample of invoices, is able to cope with all of the variabilities I have described. If the address is in a different place it will be found; if there is usually a VAT line and it's not there then that doesn't matter; if the date is in a different format it will be recognised (and converted into the standard format).

Interestingly, the AI works in the opposite way to the template approach—as more and more 'versions' of the document are found, rather than becoming more uncertain and having to create more and more templates, the AI become more confident and is able to cope with even more variability. This is because it is essentially matching the 'patterns' of one document with the learned patterns of the documents used in training.

In the case of unstructured text such as a free-form email, the AI can do two things. The first is to categorise the text by matching the patterns of words with what it has already learnt. For example, if you gave the AI a random news article it could work out whether that article was about politics, business, sport and so on, as long as it had been trained on many other tagged news articles. It would do more than simply look for the word 'football' to identify a sports article (there are many business articles about football, for example) but would instead look at the article holistically, creating an algorithmic model that represents 'sports articles'. Any new article with a similar model, or pattern, would likely be about sports.

The second main thing that an AI search capability can do is to extract 'named entities'. A named entity could be a Proper Noun—a place or a person, for example—or even a date or quantity. So, in the passage of text, "Andrew Burgess, who lives in London, wrote his second book in 2017 which was published by Palgrave Macmillan", the named entities would be 'Andrew Burgess', 'London', 'second', '2017' and 'Palgrave Macmillan'.

There are specific Named Entity Recognition (NER) algorithms that carry out this task, but they all need to be trained and tuned to make them as accurate as possible. The best NER systems for English currently produce near-human performance. Accuracy for NER can be improved by training them on specific domains of information, so one could be trained on legal documents and another on medical documents, for example.

The categorisation and entity extraction tasks can be combined to enable free-form, that is, unstructured, text to be 'read' and categorised, and then have all the meta-data extracted. This would mean that, for example, a customer who emailed a request to a company could have that email categorised so that it can be automatically forwarded to the right person in the organisation along with all the relevant meta-data that has been extracted. This meta-data can be automatically entered into the company's Case Management System so that the customer service agent has all the information available when the case is received.

Search, or Information Extraction, is probably one of the most developed capabilities in the Framework. There are established software vendors with relatively mature products, as well as many start-ups all busy building applications in this space. As you will find in subsequent chapters, the principal attraction of this capability right now is that it provides a useful complement to RPA; the robots need structured data as their inputs and AI Search can turn unstructured text into structured data, thus opening up many more processes that can be automated through RPA.

Clustering

All the capabilities I have described so far work on transforming unstructured data (images, sounds, text) into structured data. In contrast, Clustering, our fourth capability, works on structured data, and looks for patterns and clusters of similar data, within that data; that is, it is a 'classifier'. The other aspects of this capability that are different from the others is that it can (but by no means has to) learn unsupervised, and it takes a very statistical approach. But just like the others, it still requires large amounts of data to be truly valuable.

Clustering is usually the first part of a series of stages that usually end with a prediction, for example extracting insights from new data based on alignment with the original patterns, or identifying anomalous new data where it doesn't match the expected patterns. In order to be able to make those predictions or identify the anomalies, the patterns/clusters must first be discovered.

In its simplest form, this type of AI uses statistical methods to find the 'line of best fit' for all the data (in mathematical terms it is minimising the square of the distance from all the points to the line). It is then just a case of making those statistical methods more and more complex in order to cope with more and more features of the data. If there isn't enough data, then the solution can suffer from what is called 'over-fitting'—this is where a theoretical line of best fit is calculated but it bears little resemblance to any real trends in the data. The more data you have, the more confident you can be in the patterns that are found.

A good example of the use of clustering is to be able to identify similar groups of consumers within customer buying-behaviour data. Humans will usually be able to identify patterns within small data sets and will often use past experience to help them shape those patterns. But where there are thousands or millions of data points with multiple characteristics and features, humans will find it impossible to process all that information, and the AI will come into its own.

The benefit of the approach, apart from the ability to use sheer computing power to analyse data quickly, is that the AI is effectively naive to the data. It is only looking for patterns in numbers, and those numbers could relate to anything, including a person's height, their salary, their eye colour, their postcode, their gender, likes, dislikes and previous buying history. If you were in charge of the loyalty card data for a retailer, you may not have realised that there is a correlation between eye colour and propensity to buy yoghurt (I'm making this up) but, if it exists, the AI will find it. No one would have thought of trying to match these two features and so, if they were using a traditional

Business Information tool, wouldn't have thought to ask the question in the first place. It is this insight that AI can bring that provides much of its value.

Just like the other capabilities in the 'Capturing Information' group, the Clustering approaches are reasonably mature and are being used in business—it is the basis of the field of Predictive Analytics. You only have to buy anything online to see how you have been pattern-matched against other consumers and then presented with an offer to buy something that they have bought in the past. Or you may have received a special offer from your mobile phone provider because some of your usage patterns and behaviours have indicated you may be a 'flight risk' and they want to keep you as a customer. Or you may have received a call from your bank because of some of your spending behaviour doesn't fit into the normal patterns raising the possibility that you may have had your credit card skimmed.

It is this Clustering capability that benefits more than any of the others from the sheer propensity of data in the world today.

Natural Language Understanding

Now we are ready to move on to the next type of AI objective—'determining what is happening'—and its first capability, Natural Language Understanding, or NLU.

NLU is an important part of the AI world because it provides some meaning to all the text we are bombarded with, without having to resort to humans to read it all. It acts as a 'translator' between humans and machines, with the machine having to do the hard work. NLU is closely related to the Search capability I discussed earlier: some people would actually group Search under the wider banner of NLP, but I prefer to split them out as they are generally used for different objectives in business. The same can be said of Speech Recognition, but again, using these capabilities in the context of the business setting, they provide different objectives. NLU is still, to a certain degree, turning the unstructured data of a sentence into the structured data of an intent, but its primary purposes are to work out what is the right structure (syntactic analysis) and what is the meaning (semantic analysis) of the words and sentences.

NLU has a special place in the history of AI because it forms the basis of the Turing Test. This is the test that English polymath Alan Turing devised in the 1950s to define whether a machine could be classed as artificially intelligent or not. In the test, an evaluator would have a typed conversation with a computer and a human, both hidden behind screens—the computer would pass the test if the evaluator could not tell which of the conversations was with a computer or a human. The Turing Test is now an object of some debate—it is not only highly influential, particularly when discussing the philosophy of AI, but also highly criticised. A number of systems, most notably 'Eugene Goostman', a Russian chatbot, have claimed to have passed the test, but the real question, now that AI has developed so much further, is whether it is a valid test of intelligence in the first place.

NLU uses supervised learning with machine learning in order to create a model of the input text. This model is probabilistic, meaning that it can make 'softer' decisions on what the words mean. NLU is a very complex research field and I won't begin to go into the details here, suffice it to say that there are many challenges, such as being able to cope with synonymy (different words that have similar meanings) and polysemy (words that have several meanings). The phrase 'Time flies like an arrow. Fruit flies like a banana.' perfectly sums up the challenge that NLU researchers face.

We all see NLU in action most days. Siri or Cortana or Alexa all have usable natural language interfaces. The majority of the time they will understand the words you are saying (using Speech Recognition) and convert that into an intent. When these systems don't perform well it is either because the words weren't heard correctly in the first place (this is the most common reason) or the question that has been asked doesn't make sense. They will be able to understand different versions of the same question ("What was the football score?", "Who won the football match?", "Please can you tell me the result of the football?", etc.) but if you ask "Score?" they will struggle to come up with a relevant answer. Most systems will have default answers if they can't work out what you want.

Chatbots, where the questions and answers are typed rather than spoken, use NLU but without the 'risk' of Speech Recognition getting the word interpretation wrong. They therefore tend to be the simpler and preferred choice when businesses are looking to create NLU interfaces with their customers, but the use of speech recognition shouldn't be discounted if the environment and benefits are good enough.

And, of course, these personal assistants can talk or type back to you, but the answers tend to be stock phrases. For more bespoke responses a specific sub-set of NLP called Natural Language Generation (NLG) is required. NLG, which can be considered as the 'opposite' of NLU, is probably the hardest part of NLP, which is why it has only recently become commercially available. Rather than the limited set of response phrases used in the majority of chatbots, creating whole new phrases and even articles is possible with NLG. Applications include creating hyper-local weather forecasts from weather data, or financial reports from company's earnings data or stock market data. Each of these provides a short narrative in a natural language that can be hard to differentiate from one generated by a person.

NLU is also used to try and understand the emotion behind the sentences, an area known as sentiment analysis. In its simplest form sentiment analysis looks for 'polarity', that is, is the text positive, negative or neutral. Beyond that it will look for the type of emotion that is being expressed—is the person happy, sad, calm or angry, for example. The AI will, of course, look for specific words within the text but it will try and put these in context, as well as try to identify sarcasm and other tricky idiosyncrasies of language.

Sentiment analysis is already being used widely to assess online text from customers, such as through Twitter and TripAdvisor. This allows businesses to evaluate how their products or services are being perceived in the market without the expense of focus groups or surveys, and allows them to respond to potential issues in real time. Specific models, with the associated terminology and jargon, are usually generated for particular needs; understanding how customers perceive a hotel room will be quite different to how they feel about their mobile phone.

Another common application for natural language technologies is machine translation. This is where a computer will translate a phrase from one language to another. It is a complex problem to solve but recent developments in deep learning have made this much more reliable and usable. The key to success, as with most AI challenges, is the availability of data, and companies like Google have used the transcriptions from the European Parliament, where the proceedings are translated by humans between 24 different languages, as its training set. That means that translating from, say, English to French is now highly accurate, but for less common languages, such as Khmer, the official language of Cambodia, translations are generally done through an intermediary stage (usually the English language). Chinese is notoriously difficult for machines to translate, mainly because it is more of a challenge to edit sentences as a whole—one must sometimes edit arbitrary sets of characters, leading to incorrect outcomes.

The natural evolution of machine translation is real-time translation of speech (epitomised by the Babel Fish in *The Hitchhiker's Guide to the Galaxy*). Already smartphone apps are able to read and translate signs (using Image Recognition, NLU and machine translation) and can act as interpreters between two people talking to each other in different languages. Currently these capabilities are less reliable than typed translations and only work for common language pairs, but the potential uses for this technology in a business environment are huge.

NLU is leading the charge for AI in business: it allows us to communicate with computers in the way we feel most comfortable with—we don't all have to be computer wizards to be able to work them effectively. But NLU is inherently a difficult technical challenge and will always be compared with, and evaluated against, humans (unlike other AI capabilities such as Clustering which can easily surpass human performance). Therefore, until it can be said to be indistinguishable from us (i.e. truly pass the Turing Test) NLU will always be open to criticism. Applied in the right environment and with the right processes though, it can deliver substantial benefits to businesses.

Optimisation

In all the capabilities I have discussed so far, we have been manipulating data to transform it from one form to another (images to descriptions, sounds to words, words to meanings, text to information and big data to meta-data). But even though the transformed data is much more useful to us than the source data was, we haven't yet done anything really meaningful with it. That's where the AI capability of Optimisation comes in.

Optimisation is at the heart of what people generally think AI does. It is the closest analogy we get to mimicking human thought processes without having to call on true cognitive Understanding.

I use the title 'Optimisation' for brevity, but it actually includes problem solving and planning, which makes it quite a broad subject that has a lot of science behind it all. A broad definition of this capability would be that if you know a set of possible initial states as well as your desired goal and a description of all the possible actions to get you there, then the AI can define a solution that will achieve the goal using an optimum sequence of actions from any of the initial states.

Historically, as I discussed at the start of Chap. 2, optimisation and problem solving were achieved through 'expert systems', which were really nothing more than decision trees which had to be designed and configured up-front by humans. With the advent of machine learning, much of that design and configuration is carried out by the AI itself using an extreme version of trial-and-error. (There is still, by the way, a place for knowledge-based AI systems in the world today, as I shall discuss later.)

A useful way to understand the Optimisation capability is by looking at where AIs are taught to play computer games. Scientists and researchers use computer games as a kind of test and benchmark for how clever their systems are: can this particular AI beat the best human at playing that particular game? But it also provides a very useful way of explaining what the different flavours are.

So, for example, to play chess and beat Gary Kasparov required brute force—these systems 'simply' analysed every possible move much further ahead than a human can do. That is more logic than it is AI. To learn and win at Space Invaders or Breakout, however, needs DNNs and what is called 'reinforcement learning' (more of which later). From only being given the objective to maximise the score, and no other information on how to play, these AIs perceive the state of the screen, trying lots of different approaches until the score increases (shooting aliens, not getting bombed, etc.). Then, through reinforcement, they start to develop strategies (such as getting the ball behind the wall in Breakout) that get higher scores. It's a great example of self-learning, but in a relatively simple environment.

In order to learn the Chinese game of Go, things are a lot trickier. Until very recently, a computer being able to play Go was a Holy Grail for AI researchers, mainly because the game relies more on 'intuition' than logic, but also because the number of possible move combinations is trillions of times greater than a game of chess (there are more possible Go moves than there are atoms in the universe). In 2016, an AI designed by DeepMind, a British company owned by Google, beat the best Go player in Europe four games to one. AlphaGo, as the system is called, uses the same concept of reinforcement learning, this time based on studying 30 million moves from human Go matches. It then plays different versions of itself thousands of times to work out the best strategies, which it uses to create long-range plans for the actual games. Apparently, during the second match, the human player, Lee Sedol, 'never felt once in control' (a general fear of AI that is shared by many people, it must be admitted). Since then, AlphaGo has beaten the world's best Go player, Ke Jie, three games to zero.

So, the key characteristic of the Optimisation capability is that there is a goal to be achieved—an idea that needs to be reasoned, a problem to be solved or a plan to be made. At its simplest form this is done through iterative trial-and-error—small changes are made to the environment or small actions are taken, after which the situation is evaluated to see whether it is closer to the goal or not. If it is closer it carries on and makes further changes, if not it tries something different.

As I've hinted with the gaming examples above, there are some subtleties to this overly simplified description, which all come together to provide the AI Optimisation capability. I've described a number of the most common approaches below, most of which are inherently inter-connected.

Cognitive reasoning systems are the modern equivalent of the old Expert Systems (which is why some people are reluctant to describe them as true AI today). They work by creating a 'knowledge map' of the specific domain that is being modelled, which means that they do not require any data to be trained on, just access to human subject matter experts. These knowledge maps connect concepts (e.g. food and person) with instances (e.g. chips and Andrew Burgess) and relationships (e.g. favourite food). Different relationships can have different weights or probabilities, depending on their likelihood, which means that the system can be interrogated to make recommendations. Unlike decision trees, the system can start from any point on the map as long as a goal has been defined ('what is Andrew's favourite food?', 'which person likes chips?') and can handle much greater complexity: its 'clever' bit is in the way it comes to a recommendation using the shortest route through the map rather than asking a linear series of questions. Compared to machine learning approaches, cognitive reasoning systems have the major advantage that the recommendation that has been made is fully traceable—there is none of the 'black box' characteristics that machine learning systems suffer from. This makes these sorts of systems ideal for regulated industries that 'need to show their working' (Fig. 3.2).

Beyond Cognitive Reasoning, the majority of AI optimisation approaches are based around algorithms. A key AI concept involves the idea of breaking a big problem down into much smaller problems and then remembering the solutions for each one. This approach, known as Dynamic Programming, is

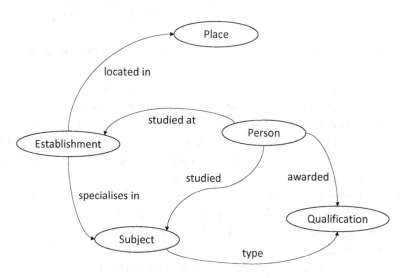

Fig. 3.2 Knowledge map

epitomised in the Coin Change Problem, which asks: how can a given amount of money be made with the least number of coins of given denominations? If my coins are in units of 1,4,5,15,20 and the total amount I need to make is 23, then a system which looked at it as a series of independent, 'blind' decisions would start with the biggest coin first and then add smaller coins until the total was reached: this would be 20+1+1+1, so four coins. A dynamic system though would have broken the problem down into smaller problems, and stored each of the optimal solutions. It would have come up with an answer of 15+4+4, so just three coins.

The system works by looking ahead at lots of different ways that the end objective could turn out to be based on, analysing a large sample of the different individual steps that can be taken. Those sample actions that lead to, or get close to, the desired goal are remembered as favourable and therefore more likely to be chosen.

A common AI approach that exploits this method is called the Monte Carlo Tree Search (MCTS). When, for example, playing a game it can 'play out' lots of different versions of potential moves of both players, with each move creating a branching tree of child moves. Using back propagation (a form of feedback loop), those moves that achieve the desired goal have the connections between each of the nodes that led to that result strengthened so that they are more likely to be played.

These MCTS-like approaches do have a couple of drawbacks: they need to balance the efficiency of a small sample size of moves with the breadth of covering as many moves as possible (this leads to some randomisation being built into the choice of child nodes), and they can be slow to converge on to the optimal solution.

Reinforcement Learning, an area of AI that is seeing a great deal of activity at the moment, helps to alleviate some of these inefficiencies. Reinforcement learning is really an extension of dynamic programming (it is sometimes referred to as 'approximate dynamic programming') and is used where the problems are much more complex than, for example, the coin change problem or simple board games, and where there are more unknowns about the states of each move that is taken. Reinforcement learning uses extreme trial-and-error to update its 'experience' and can then use that machine-learned experience to determine the most optimal step to take next, that is, the one that will get it closer to achieving the goal. This approach is slightly different to the supervised learning approaches I described in Chap. 1 in that the 'right answer' is never given in the training, nor are errors explicitly corrected: the machine learns all of this itself through the trial-and-error iterations.

A tactic that is usually employed in reinforcement learning is to have the AI systems play against each other. DeepMind's AlphaGo, once it had been trained on human games, then played itself (or, to be exact, two slightly different versions of AlphaGo played each other) over thousands of games in order to refine its game even further. If it had only learnt from human games it could only ever be as good as humans, whereas if it plays itself, it is possible for it to surpass human skill levels. This combative approach, known as Generative Adversarial Networks, as well as allowing the algorithms to tune themselves, is a great way to run simulations to test the viability of AI systems and to run different war-gaming scenarios.

Another 'weakness' in AI Optimisation approaches has been that the systems tend to look at the short-term gains at the expense of the longer-term strategy. Recent advances in AI have seen a number of different systems combined to provide both these perspectives, especially where the domain is particularly complex: a 'policy' algorithm will look at the next best move, whilst a 'value' algorithm will look at how the problem, or game, might finish up. The two algorithms can then work together to provide the best outcome.

Facebook has already been able to train chatbot agents to negotiate for simple items, in some cases doing it as well as humans. Facebook Artificial Intelligence Research first used supervised learning to train the agents on a large number of human-to-human negotiation scripts. This stage helped them imitate the actions of the humans (mapping between language and meaning), but didn't explicitly help them achieve their objectives. For this they used reinforcement learning, where two AI agents would 'practise' negotiating with each other (interestingly they had to fix the language model in place at this stage because they found that, if the agents were allowed to continue learning the language elements, they would start to create their own private language between them). At the end of each negotiation, the agent would be 'rewarded' based on the deal it had managed to achieve. This was then fed back through the model to capture the learning, and make the agent a better negotiator. Following training, it could then negotiate for the same types of items with humans, matching the capabilities of other human negotiators.

Similar ideas of combining a number of AI techniques were used when an AI system, called Libratus, was able to beat experienced players of poker—this system used three different types of AI: the first relied on reinforcement learning to teach itself the game of poker from scratch; a second system focused on the end game, leaving the first system to concentrate on the immediate next moves; and, because some of the human players were able to detect trends in how the machine bet, a third system looked for these patterns overnight and introduced additional randomness to hide its tracks.

Optimisation AI can be applied to many situations where there is a specific goal to be achieved, such as winning a hand of poker or negotiating for items. It can, as I mentioned earlier, provide performance that exceeds human capabilities. Other typical uses for the Optimisation capability include route planning, designing rota for shift workers (such as nurses) and making recommendations.

All the approaches described in this section are just a sample of how AI can attempt to solve problems, but should give you a feel for the general strategies that can be employed. At the heart of it is the idea that big decisions can be broken down into many small decisions, and that these can then be optimised using trial-and-error so that a defined goal can be achieved, or a specific reward maximised.

Prediction

Prediction employs one of the core ideas of AI in that it uses lots of historical data in order to match a new piece of data to an identified group. Thus, prediction generally follows on from the Clustering capability described earlier in the chapter.

I've already mentioned one of the more common uses of prediction, that of recommending related online purchases ('you bought this book therefore you will probably like this other book'). So, in this case, what the retailer calls a 'recommendation' is actually a prediction they make in order to sell you more stuff.

Some decisions can be described as predictions—if you apply for a loan that is assessed by a machine, then the AI will try and predict whether you will default on the loan or not. It will try and match your profile (age, salary, regular spending, other loans, etc.) with other customers that have similar profiles. If those other matching customers generally default on their loans then you will likely be refused the credit line.

Prediction can also follow on from the Search capability. Because Search can look for patterns within text, it can match these with patterns from some pre-defined requirements. For example, CVs (aka résumés) can be matched with job descriptions in order to predict good candidates for the role.

The prediction capability differs from Optimisation in that it doesn't have a specific goal to achieve. There are no steps to determine to achieve a specific objective—we are 'just' matching a new data point with our historical data.

At a simple level, predictions can be made from just a few 'features'—these are the specific characteristics that are being measured, such as the number of bedrooms in a house and the size of the garden. If you had a table of this data and the actual price of each of the houses, then you could use this capability to predict the price of another house if you knew the number of bedrooms and size of the garden. And you wouldn't necessarily need AI to do this, or even a computer.

But there are generally more factors (or features) which determine the price of a house than just these two things. You might also consider how many stories it has, whether it is detached, semi-detached or terraced, whether it has parking, a utility room, a swimming pool, its postcode and so on. All this makes it much more difficult to predict the house value without using some AI 'magic'.

Clearly, the more features that are considered, the more training data needs to be used, so that as many 'variations' of all the features can be captured and built into the model. This model will have considered all the different influences (or 'weights') that each of the features has on the house price. By entering in the values of a selection of features of a different house, the model fits (using Regression Analysis in this case) those features as closely as possible and can therefore predict the house price. It is normal for the system to give a figure for its level of confidence (i.e. the probability that it is correct).

Once the number of features is in hundreds or thousands then things become more complex—this just means that more algorithms, more training data and more computing power are needed. At the extreme end of this, the systems that are used to predict our weather are some of the most powerful in the world.

An important aspect of the Prediction capability is to remember that it is essentially naive. By this I mean that it is only manipulating numbers and has no underlying understanding of what that data actually means. Those house features could be car features or weather features or people features—to the computer they are just numbers.

This naivety, whilst seeming like a virtue, is also one of Prediction's biggest challenges: that of unintended bias. Normally AI is lauded because it is not influenced by the prejudices that humans inherently suffer from (there are many studies that show that, although people may claim not to be biased in, say, recruiting staff, there is usually some sort of unconscious bias involved). But the unbiasedness of AI is only as good as the data that is used to train it—if the training set contains CVs and hiring decisions that have been made by humans, the biases that were in the original data will be 'trained into' the AI.

Another key issue for AI Prediction is the opaqueness of the decision-making process. When an AI makes a prediction, for example that a person is more than likely to default on a loan and therefore their application should be rejected, that prediction is based on all the training data. The training data has been used by the machine to build an algorithmic model that it then uses to predict new cases from. For the more complex algorithmic models this is just a matrix of numbers that makes no sense to a human trying to read it, and therefore the reason (or, more likely, reasons) that the person had their loan refused is not easily available to scrutinise. (Simpler algorithmic models, such as Classification and Regression Trees, do provide some transparency and are therefore more popular). Of course, we don't need to understand the workings of every prediction that is made (it would be interesting to know why the house price predictor came up with that valuation, but it is the actual house price we are most concerned with), but some industries, especially those that are regulated, will require it. Also, if you are the person that has been refused the loan, you should have the right to know why.

AI is being used to predict many things, including, in some US courts, a defendant's risk profile. When the consequences of those predictions could have an influence on whether someone is found innocent or guilty, then the challenges I have described above become very serious indeed. As Melvin Kranzberg's first law of technology succinctly states: technology is neither good nor bad; nor is it neutral. These challenges of 'algorithmic transparency', naivety and unintended bias will be discussed in more detail in Chap. 8.

AI Prediction is one of the most active areas in the field at the moment. Where there are lots of good data, predictions can generally be made from that data. That doesn't necessarily mean that predictions need to be made or that they will be useful in any way, but there are lots of cases where this can be very beneficial indeed, including predicting valuations, yields, customer churn, preventative maintenance requirements and demand for a product.

Understanding

I include this section on AI Understanding in the book only as a way to describe what is *not* currently available in the AI world to businesses, or to anyone outside of a research lab. It should be seen as a countermeasure to all of the hype that can be put out by over-excited marketing departments.

By 'understanding', I am generally referring to the ability of a machine to have conscious awareness of what it is doing or thinking (or to act like it does—see next paragraph). This implies it can understand the intent and motivations of people, rather than just blindly crunching numbers about them. It is usually described by the concept of Artificial General Intelligence (AGI), where the AI is able to mimic all of the capabilities of the human brain, not just the very narrow ones we have discussed so far.

There is an interesting subtlety to the description of AGI. John Searle, a philosopher, differentiated 'strong AI' and 'weak AI'. A strong AI system can think and have a mind, whereas a weak AI system can (only) act like it thinks and has a mind. The former assumes that there is something special about the machine that goes beyond the abilities that we can test for. Ray Kurzweil, the futurologist, simply describes strong AI as when the computer acts as though it has a mind, irrespective of whether it actually does. For our purposes, we will stick to this definition as we are dealing with practicalities rather than philosophical debate.

So, what sort of tests would we use to claim strong AI? I have already mentioned the Turing Test as a (limited) test of AI. Other tests that people have proposed are:

- The Coffee Test (Wozniak)—A machine is given the task of going into an average home and figuring out how to make coffee. It has to find the coffee machine, find the coffee, add water, find a mug and brew the coffee by pushing the proper buttons.
- The Robot College Student Test (Goertzel)—A machine is given the task of enrolling in a university, taking and passing the same classes that humans would, and obtaining a degree.
- The Employment Test (Nilsson)—A machine is given the task of working an economically important job, and must perform as well or better than the level that humans perform at in the same job.

No AI systems are anywhere near passing these tests, and any systems that are even close will be a combination of many different types of AI. As I've described in the previous sections of this chapter, there are a number of different types of AI capabilities, each one very specialised in what it does. This means, for example, an AI capability that is being used to recognise images will be useless at processing language. This is the concept of Artificial Narrow Intelligence (ANI). Even within the capability groups I have described there is little or no crossover between specific uses—if I have a system that extracts data from invoices, it will not be able to do the same for remittance advices, without training the system from scratch. The same could even be true if I wanted to take a trained invoice extractor from one business to the another— there may be enough variation between businesses to mean that retraining would be required.

Where our brain is so much cleverer than AI is where it is able to use different cognitive approaches and techniques in different situations and, importantly, take learnings from one situation and apply them to a completely different one. For example, I may know that the value of a house generally increases with the number of bedrooms it has, but I can then apply the same concept to other objects (computers with more hard drive capacity cost more) but also know that this shouldn't be applied to everything (cars with more wheels are generally not more expensive). AI is not able to do this right now.

There is work being done to try and create AGI. At a national scale, there is the Blue Brain Project in Switzerland (which aims to create a digital reconstruction of the brain by reverse-engineering mammalian brain circuitry), and the BRAIN Project in the United States, which is also looking to model real brains. Organisations such as OpenCog (an open-source AGI research platform), the Redwood Center for Theoretical Neuroscience and the Machine Intelligence Research Institute are all at the forefront of researching various aspects of AGI.

Interestingly, there has been some progress in getting neural networks to remember what they have previously been taught. This means they could, in theory, be able to use learnings from one task and apply it to a second task. Although that may sound like a simple thing for a human to do, 'catastrophic forgetting' (when new tasks are introduced, new adaptations overwrite the knowledge that the system had previously acquired) is an inherent flaw in neural networks. DeepMind, the UK AI company owned by Google, is developing an approach it calls Elastic Weight Consolidation which allows the algorithm to learn a new task whilst retaining some of the knowledge from learning a previous task (they actually use different Atari computer games as their test tasks). This is showing promise but is a long way from being put to practical use.

Despite the research and small steps that are being taken, the ability for a computer to fundamentally understand what it is doing is still a long way off (some argue that it could never be possible). Even using Searle's definition of weak AI, there are major hurdles to overcome. Some of these are very technical (such as the challenge of catastrophic forgetting) and some are simply down to the fact that the necessary computing power doesn't exist at the moment. But, as described throughout this book, there are huge advancements being made in ANI that provide real benefits to people and businesses. Some of these make our lives simpler whilst others are able to far exceed our own competencies in specific tasks. By understanding each of these capabilities, and their respective limitations, we are able to benefit from AI technologies today and tomorrow.

Using the AI Framework

The AI Capability Framework is my attempt to bring some clarity and order to the diverse, complex and often confusing field of AI. By 'boiling it down' into a set of discrete capabilities, AI hopefully becomes accessible to those who want to benefit from the technology but do not have the skills, experience (or desire) to understand the technical aspects beyond the high-level appreciation I have provided in this chapter.

I have tried to differentiate each of the capabilities as much as possible, but there are inevitably some overlaps between each of them, for example between speech recognition and NLU, and clustering and prediction. The boundaries between these are blurry; some people would describe both speech recognition and NLU as a sub-category of NLP but I think they sit better as discrete capabilities; and some people might separate out planning and optimisation, but I think they are close enough aligned to keep them as one. So, please don't get too hung up on some of the nuances—AI is a complex subject, with many different viewpoints and opinions, and is constantly changing. The framework should be treated as your guide rather than a technical manual (Fig. 3.3).

So, the knowledge you now have should enable you to do three things:

- **Identify the right AI capabilities for your business need**. What are your business objectives and could AI provide all or part of the solution? Are you looking to capture information or understand what is happening, or both? Do you want to replace existing capability (computer or human) with AI, or do you want to augment their capabilities further? Which specific capabilities will you need to create a solution? Will that require a supervised learning approach, and what data do I have available?

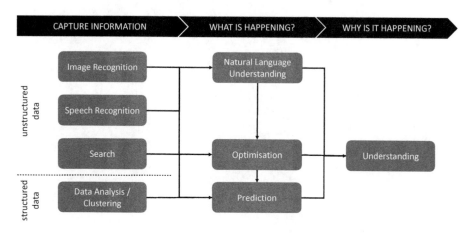

Fig. 3.3 The AI framework

- **Look beyond the hype**. What capabilities do the AI vendors actually have, rather than those they claim they have? How do these capabilities match my requirements? Do their assertions about their product make sense? Are there gaps that will need to be filled with different AI solutions or traditional technology? Will it be as easy to implement as they claim?
- **Be realistic**. What are the limitations of the AI capabilities that I might need? Is AI the most appropriate solution for this or are there simpler or more effective solutions? Do I have the necessary data available to train the system adequately? Will I need to bring in external support to help me get going?

Chapter 5 provides real examples of the different AI capabilities being used in businesses today. They are focused on common strategic objectives (enhancing customer service, optimising business processes and generating insights) and combine different capabilities to achieve these. Each of the examples references the capabilities that are being used and, where appropriate, what challenges and limitations were experienced by those responsible for delivering the solutions.

Some of the approaches and examples that will be described required additional technologies (such as cloud computing and RPA) to enable them to extract their full value. The following chapter outlines what these associated technologies are and why they are such a good fit with AI.

The AI Start-Up's View

This is an extract from an interview with Vasilis Tsolis, founder and CEO of Cognitiv+, an AI start-up in the legal sector.

AB   What got you into the world of AI in the first place?

VT   My personal background is not an obvious path for a co-founder of an AI start-up. I am a chartered civil engineer who switched my career by undertaking a law degree. After graduation, my career was moving between law, engineering and commercial management in various sectors including infrastructure, construction and energy. After a few years, something became very clear to me: people spend substantial amounts of time reading contracts, and most of the time there is a tremendous repetition of those tasks. And that is where the problem lies: rather than focus on advanced solution discovery, professionals consume their daily lives with data gathering. The solution was an obvious one; using AI will free the professionals to focus on what they are good at. That is why we started Cognitiv+ in 2015, which has been a fantastic trip so far.

AB   What is it like building an AI start up at the moment?

VT   This is a period where people apply AI or attempt to use AI in almost all aspects of our lives, where creativity sometimes trumps technology. There are major projects in process, such as autonomous cars and it very clear now to everyone that AI is here to stay. It is a very exciting period, but there are quite a lot of challenges towards the road of maturity.

Working with a number of professional clients, we see numerous opportunities and dozens of use cases, some possible now, some reachable within the mid-term future, some look close to science fiction.

Technology advancements advertise extraordinary abilities and this goes throughout the food chain, where clients are constantly bombarded with innovative ideas where the success rate is variable. This could be because they are early in their roadmap and suffer teething problems, or because of a lack of technical or subject matter depth which is sometimes not an easy fix.

As with any early adoption, it is the clients and investors that get the pressure to isolate the true technologists and innovators with clear business skills, and some in-house knowledge on how technology works will definitely bring certainty and a quick return on investment.

AB   So what value do your customers get out of your software?

VT   Reading text takes time, even Romans were complaining about this and AI can help that. Cognitiv+ targets legal text, contracts and the regulation corpus.

Using AI should be seen as an extrapolation of automation, a technology that improves our lives, makes things faster, simpler and enables us to explore tasks in a way that was not possible before. For example, clients can use our software to analyse contractual risk on their whole portfolio within hours in a way that was never possible before. Time-consuming tasks can be delegated and free up professionals' schedules, where technology can provide them with holistic views of their third-party risk to levels that have been unprecedented.

AB What do customers or potential customers need to focus on if they are going to get the maximum value from AI?

VT While many companies hit the ground by starting to write code, there is a lot of preparatory work that projects need to undertake and ensure success.

The first focus will be the data. Data scientists require massive data sets and the new techniques, supported by hardware innovation, command new levels of data quantities. The good news is that we all produce more data every day, but this is not necessarily for all the aspects of professional services. Some data custodians can move faster to this gold rush because they have better access to data from others.

Quantity is not enough though—it is also the quality that will determine if the data can be used. But this is a synthetic challenge that can be tackled only if it is isolated and understood on a case-by-case basis.

Second, teams need to make an early decision on what success would look like, making business and technology targets that are reachable.

The fact that algorithms provide answers that are probabilistic and not deterministic makes the way we manage AI projects very different from any previous IT transformation projects companies have participated in.

Thirdly, an important focus would be the talent and skills diversity that a team has. A sensible balance between business subject matter experts, data scientists and coders would determine how successful a project will be.

AB How do you see the market developing over the next 5 years or so?

VT While I can predict where we are heading in the next 5 months, it is practically impossible to hit anything close to a 5 year prediction. The reason is simple—all the parameters are rapidly changing: the type of data we are surrounded with, the size of data sets, the maturity of the algorithms, hardware improvement, the list goes on.

But here is one prediction: as we have started using various NLP, NLU and machine learning techniques to interpret text and documents, it seems that certain types of text, certain sources, can be analysed better than others. People will pick this up and will write in a way that can be processed and summarised by machines. Why do that? It is for the same reason where we think twice what tags we are using for our blog articles as this will enable Search Engine Optimisation (SEO) bots to categorise our text accordingly and disseminate it to the right channels.

Typically, we write in a way that our audience will understand us—my take is that we will continue to do that—the difference is that in the crowd or readers will also be some bots. What might this look like? Perhaps, simpler and shorter phrases where the object and subjects are clear and NLP algorithms can consume them much better?

4

Associated Technologies

Introduction

Artificial Intelligence can do a lot of things, but it can't do everything. Quite a lot of the time implementing a stand-alone AI solution will satisfy whatever objectives you are looking to achieve. Sometimes AI will rely on other technologies to make it work well, and other times it will complement them so that they both work better. This chapter covers some of the associated technologies that anybody looking to implement AI will need to consider.

Some of these technologies are software-based, such as RPA. This is a relatively new technology that automates rules-based processes but struggles to handle unstructured data and any decision making. Cloud Computing, a combination of software and hardware capabilities, is a key enabler of AI, and many of the AI applications that we are seeing now would not be possible without it.

There are also some hardware technologies that enable, and can be exploited by, AI. Physical robots (as opposed to the software-based RPA) can be made more intelligent with the application of AI, and the Internet of Things (IoT) can provide very useful data sources for AI systems. One of the technologies that I have included in this chapter isn't really a technology at all, but is powered by human beings: Crowd Sourcing enables data to be tagged and cleaned in an efficient and flexible way, and therefore is extremely useful to AI developers.

I have not included more common enterprise systems, such as Enterprise Resource Planning (ERP) and Customer Relationship Management (CRM) systems, in this summary. All could be considered sources of data, and some

© The Author(s) 2018
A. Burgess, *The Executive Guide to Artificial Intelligence*,
https://doi.org/10.1007/978-3-319-63820-1_4

claim to have AI capabilities built into them (e.g. email systems can be a good data source and also use AI to identify spam messages), but overall, they have only general associations with AI.

AI and Cloud

Cloud Computing is where a network of remote servers, hosted on the Internet, are used to store, manage and process data, rather than the 'traditional' approach of using a local (on-premises) server or a personal computer to do those tasks.

Because of these abilities to store, manage and process data away from the user's device (PC, mobile 'phone, etc.) and instead on high-performance, high-capacity specialised servers, Cloud Computing has become almost an integral part of how AI systems operate today. As cloud technology itself matures, the two will become inextricably linked—many people already talk of 'Cloud AI' as the Next Big Thing.

One of the more straightforward applications of cloud combined with AI is the making available of large, public data sets. Most AI developers, unless they are working for large corporates, don't have their own data sets to train their systems, but instead rely on these public databases. As I mentioned in the Big Data section of Chap. 2, there exist quite a few of these datasets, covering subjects such as Computer Vision, Natural Language, Speech, Recommendation Systems, Networks and Graphs, and Geospatial Data.

But the cloud provides more than just access to data—more often it will actually process the data as well. We can experience this as consumers every time we use a service like Amazon's Echo. Although there is a small amount of processing power on the device itself (mainly to recognise the 'wake word'), the processing of the words into meaning (using Speech Recognition and NLU) is done by software on Amazon's own servers. And, of course, executing the actual command is also done on the cloud, perhaps sending the instruction back to your home to, say, turn the kitchen light on.

On an enterprise scale, even more of the processing work can be done on the cloud, on huge server farms full of powerful machines. The biggest challenge to this model though is that the data generally needs to sit on the cloud as well. For enterprises with huge amounts of data (perhaps petabytes' worth) then transferring all of this to the cloud can be impractical. As a solution to very long upload times, Amazon actually offers a huge truck (called the Amazon Snowmobile) that contains 100 petabytes of computer storage—the truck is driven to your data centre and plugged in so that the data can be uploaded to the truck. It then drives back to the Amazon Data Centre and downloads it all. But, as network speeds improve, there will be less and less need for these sort of physical solutions.

Another challenge to Cloud AI is that it can be perceived as risky to store data off-premises, especially if the data is confidential, such as a bank's customers' details. The security aspects can be covered by the type of cloud service that is procured—the best suppliers can now offer security provisions that are as good as, or better than, hosted or on-premises solutions. The UK Government in fact released guidance at the start of 2017 that states it is possible for public sector organisations to safely put highly personal and sensitive data into the public cloud.

The cost, though, of setting up an operation that can store, manage and process AI data effectively, securely and economically means that the market for Cloud AI is currently dominated by a small number of suppliers, namely Amazon, Google and Microsoft. These companies actual offer a complete set of AI services, including access to ready-to-use algorithms.

Generally, the AI Cloud offerings are made up of four main areas (I've used Amazon's model as a basis for this description):

- Infrastructure—this consists of all the virtual servers and GPUs (the processor chips) that are required to house the applications that train and run the AI systems
- Frameworks—these are the AI development services that are used to build bespoke AI systems, and tend to be used by researchers and data scientists. They could include pre-installed and configured frameworks such as ApacheMXNet, TensorFlow and Caffe.
- Platforms—these would be used by AI developers who have their own data sets but do not have access to algorithms. They would need to be able to deploy and manage the AI training as well as host the models.
- Services—for those who don't have access to data or algorithms, the AI Services offer pre-trained AI algorithms. This is the simplest approach for accessing specific AI capabilities with a minimum knowledge of how they work technically.

The ability to tap into relatively complex, pre-trained algorithms is a boon for anyone wanting to build AI capability into their applications. For example, if you want to build an AI application that has some NLU capability (a chatbot, for example) then you could use Amazon's Lex algorithm, Microsoft's Linguistic Analysis or Google's (wonderfully named) Parsey McParseFace. Each of these is a simple Application Programming Interface

(API), which means that they can be 'called' by sending them specific data. They will then return a result to you, which can be read by your application.

One of the interesting things about these services is that they are either free or cheap. Microsoft's Language Understanding Intelligent Service (LUIS) offering is currently based on a threshold of API calls per month—below this threshold it is free to use, and every thousand calls above that is charged at just a few cents. Other algorithms are charged as a monthly subscription.

Many businesses today are taking advantage of Cloud AI rather than building their own capabilities. For example: a brewer in Oregon, USA, is using Cloud AI to control their brewing processes; a public TV company uses it to identify and tag people that appear on its programs; schools are using it to predict student churn and an FMCG company uses it to analyse job applications.

As I've discussed in Chap. 2, one of the motivations for these companies to offer their AI technologies for virtually nothing is that they get access to more and more data, which in many ways is the currency of AI. But despite the 'greedy corporate' overtones, Cloud AI does have the feeling of democratisation about it, where more and more people have simple and cheap access to these very clever technologies.

AI and Robotic Process Automation

Robotic Process Automation (or RPA) describes a relatively new type of software that replicates the transactional, rules-based work that a human being might do.

For clarity, it is important to differentiate between RPA and the traditional IT systems. RPA—at its most basic level—utilises technology to replace a series of human actions (which is where the 'robot' terminology comes in). Correspondingly, not all technologies provide automation, and replacing a single human action with technology (e.g. a mathematical equation in a spreadsheet) is not considered RPA. Similarly, much automation is already embedded into software systems (e.g. linking customer information across finance and procurement functions), but since it is part of the normal feature functionality of a system, it is generally not considered as RPA, but simply a more powerful system(s).

In an ideal world, all transactional work processes would be carried out by large, all-encompassing IT systems, without a single human being involved. In reality, whilst many systems can automate a large part of specific processes and functions, they tend to be siloed or only deal with one part of the end-to-end process (think of an online loan application process that will have to get data from, and input data to, a web browser, a CRM system, a credit checking system, a finance system, a KYC system, an address look-up system and probably one or two spreadsheets). In addition, many businesses now have multiple systems that have been acquired as point solutions or simply through numerous mergers and acquisitions. The default 'integration' across all of these systems, tying the whole end-to-end process together, has traditionally been the human being. More often than not, these human beings are part of an outsourced service.

Robotic Process Automation can replace nearly all the transactional work that the human does, at a much lower cost (as much as 50% lower). The RPA systems replicate (using simple process mapping tools) the rules-based work that the human does (i.e. interfacing at the 'presentation layer') which means that there is no need to change any of the underlying systems. A single process could be automated and be delivering value within a matter of weeks. The 'robot' can then continue processing that activity 24 hours a day, 7 days a week, 52 weeks a year if required, with every action being completely auditable. If the process changes, the robot only needs to be retrained once, rather than having to retrain a complete team of people.

So, to give a simple example, if a law firm is managing a property portfolio on behalf of a client, they would be expected to carry out Land Registry checks at some point. This is commonly a paralegal role that might involve the person getting a request from a lawyer, or from the client directly, probably via a template, an email or a workflow system. The person would read the relevant information off the form, log into the Land Registry site, enter the information into the site and read the results that came back from the search. They would then transpose that information back onto the form and respond to the initial request. It is possible for this whole process to be handled by a software 'robot' without the need for any human intervention.

Although this is a very simple example, it does demonstrate some of the benefits of RPA:

- the cost of the robot is a fraction of the cost of the human (between a third and a tenth);
- the robot works in exactly the same way as the human would, so no IT or process changes are required;
- once trained, the robot will do the process exactly the same way 100% of the time;
- every step that the robot takes is logged, providing full auditability;

- the robot can carry out the process in the middle of the night or over a weekend if necessary and
- the robot will never be sick, need a holiday or ask for a pay rise.

What this means is that wherever there are processes that are rules-based, repeatable and use (or could use) IT systems, the person doing that process can be replaced by a software robot. Some further examples of processes that can be automated are:

- Employee on-boarding
- Invoice processing
- Payments
- Conveyancing processing
- Benefit entitlement checks
- IT Service Desk requests

These are just a small sample of the sorts of processes that can be automated through RPA. Any comprehensive review of the processes carried out in a back office environment can identify a large number of candidates for automation. Some examples of where RPA has demonstrated particular benefits include:

- O2 replacing 45 offshore employees, costing a total of $1.35m a year, with ten software robots, costing $100,000. Example processes included the provisioning of new SIM cards. The telecoms firm then spent its savings of $1.25m on hiring 12 new people to do more innovative work locally at its headquarters.
- Barclays Bank have seen a £175 million per annum reduction in bad debt provision and over 120 Full Time Equivalents (FTE) saved. Example processes include:
    - Automated Fraudulent Account Closure Process—Rapid closure of compromised accounts
    - Automated Branch Risk Monitoring Process—Collation and monitoring of branch network operational risk indicators
    - Personal Loan Application Opening—Automation of processes for new loan applications.
- Co-operative Banking Group has automated over 130 processes with robotic automation including complex CHAPs processing, VISA charge-back processing and many back office processes to support sales and general administration.

As well as delivering cost savings, RPA is having a huge impact on the way companies are organising their resources: shared service centres are ripe for large-scale automation, and outsourced processes are being brought back in-house (on-shore) because the costs and risks are much more favourable after automation. (This, by the way, is creating a huge threat to the viability of Business Process Outsourcing (BPO) providers.

In implementing RPA, there are a number of aspects that need to be considered. Firstly, there is a need to understand whether the robots will be running 'assisted' or 'unassisted'. An assisted robot generally works on parts of processes and will be triggered to run by a human. In a contact centre, for example, a customer service agent could take a call from a customer who wishes to change their address. Once the call is complete, the agent can trigger the robot to carry out the changes, which may be required across a number of different systems. Meanwhile the human agent can get on with taking another call. Unassisted robots work autonomously, triggered by a specific schedule (e.g. every Monday morning at 8) or an alert (a backlog queue is over a certain threshold). They will generally cover whole processes and are therefore more efficient than unassisted robots. Different RPA software packages are suited to these two different scenarios in different ways.

Once the type of robot has been selected, the candidate processes for automation can be considered. There are a number of characteristics that make for a good candidate process:

- Rules-based, predictable, replicable—the process needs to be mapped and configured in the RPA software; therefore, it has to be definable down to key-stroke level.
- High volume, scalable—in most cases a process that is high volume (e.g. happens many time a day) will be preferable because it will deliver a better return on the investment.
- Relies on multiple systems—RPA comes into its own when the process has to access a number of systems, as this is where people are generally employed to integrate and move data around between them.
- Where poor quality results in high risk or cost—the exception to the high-volume criteria is where a low-volume process might have a large risk associated with it, such as in making payments, and where compliance and accuracy are the main concerns.

These criteria provide plenty of opportunities for RPA in most large businesses. But they do have some limitations. The reason that RPA software is of particular interest to exponents of AI is that, although the RPA software is really clever in the way that it manages processes, the robots are effectively 'dumb'; they will do exactly what they are told to do, with unwavering compliance. In many situations that is a good thing, but there are situations where there is ambiguity, either in the information coming in, or in the need to make a judgement. This is where AI comes to the fore.

One of the biggest constraints of automating processes through RPA is that the robots require structured data as an input. This could be, for example, a spreadsheet, a web form or a database. The robot needs to know precisely where the required data is, and if it's not in the expected place then the process will come abruptly to a halt. Artificial Intelligence, and particularly the Search capability, provides the ability to have unstructured, or semi-structured, source data transformed into structured data that the robots can then process.

Examples of semi-structured data would include invoices or remittance advices—the information on the document is generally the same (supplier name, date, address, value, VAT amount, etc.) but can vary considerably in format and position on the page. As described in the previous chapter, AI Search is able to extract the meta-data from the document and paste it into the system-of-record even though every version of it may look slightly different. Once in the main system, the robots can use the data for subsequent automated processing.

The robots can even use the AI output as trigger for them to run. For example, a legal contract can be considered a semi-structured document (it has some common information such as name of parties, termination date, limits of liability, etc.). An AI Search capability can extract this meta-data for all of a business's contracts so that they can manage their total risk portfolio. RPA robots could be triggered if there was a change in regulation and all contracts of a specific type (e.g. all those under English and Welsh Law) needed to be updated.

Another area where RPA falls down is where judgement is required as part of the process. For example, in processing a loan application, much of the initial stage (such as taking the information from the applicant's web form submission and populating the CRM system, the loan system and carrying out a credit check) can be automated through RPA. But at some point there needs to be a decision made whether to approve the loan or not. If the decision is relatively simple this can still be handled by RPA—it would be a case of applying scores to specific criteria with certain weightings, and then checking whether the total score was below or above a threshold. But for more complex decisions, or where it seems like 'judgement' is required, then the AI prediction capability can be used. (This could either be through a Cognitive Reasoning engine or using a Machine Learning approach.)

Thus, using a combination of RPA and AI allows many processes to be automated end to end, which inherently provides greater efficiency benefits than would partly automated ones.

Conversely, RPA can also help AI automation efforts. As mentioned above, robots are very good at extracting and collating data from many different sources. Therefore, RPA can be used as a 'data supplier' for AI systems. This could also include manipulation of the data (e.g. remapping fields) as well as identifying any unusable (dirty) data.

So far I have mainly focused on automating business processes, but RPA can also be used to automate IT processes as well. The concept for automating IT is exactly the same as for business processes: replace human staff doing rules-based work with software agents. Many of the tasks carried out by an IT Service Desk can be automated; common examples include password resets and provisioning of additional software on a user's desktop. One user of RPA reported a reduction from 6 minutes to 50 seconds for the average incident execution time for their Service Desk.

RPA can also work autonomously on infrastructure components, triggered by system alerts: a robot could, for example, reboot a server as soon as it received an alert that the server was no longer responding. Just like business process automation, the robots can access almost any other system, and there will be no disruption or changes required to those underlying systems. RPA in this scenario can be considered a 'meta-manager' that sits across all the monitoring and management systems.

For IT automation, AI can enhance the capabilities of RPA. By using Optimisation and Prediction capabilities they can be trained from run-books and other sources, and will continue to learn from 'watching' human engineers. The AI systems can also be used to proactively monitor the IT environment and its current state in order to identify trends as well as any changes to the environment (a new virtual server, for example) and then adjust their plans accordingly.

Companies that have implemented a combination of RPA and AI solutions have seen, for example, 56% of incidents being resolved without any human intervention, and a 60% reduction in the meantime to resolution.

So, RPA provides a useful complementary technology to AI. In addition to AI enabling more processes, and more of the process, to be automated, RPA also aids the data collation efforts that AI often needs. The two technologies also work well together when there are larger transformational objectives in mind. Enabling a self-service capability in a business can be a good way to improve customer service at the same time as reducing costs. This could be achieved through a combination of AI systems managing the front-end customer engagement (e.g. through chatbots) and RPA managing the back-end processes.

AI and Robotics

One of the first practical applications of AI was in a physical robot called SHAKEY, which was designed and built by Stanford Research Institute between 1966 and 1972. SHAKEY used computer vision and NLP to receive instructions which it could then break down into discrete tasks and move around a room to complete the objective it had been set. Although we would now see it as very rudimentary, it was, at the time, at the forefront of AI research.

Today we have robot vacuum cleaners costing a few hundred pounds that can clean your house autonomously, and, although they can't respond to voice commands, this would be relatively easy to implement. Other examples of physical robots that use AI are

- **Autonomous vehicles**—driverless cars and trucks use AI to interpret all the information coming in from the vehicle's sensors (such as using computer vision to interpret the incoming LIDAR data, which is like RADAR but for light) and then plan the appropriate actions to take. All of this needs to be done in real time to ensure that the vehicle reacts quickly enough.
- **Manufacturing robots**—modern robots are much safer and easier to train because they have AI embedded in them. Baxter, a robot designed by Rodney Brooks's ReThink Robotics, can work on a production line without a cage because it has the ability to immediately stop if it is going to hit someone or something. It can be trained by simply moving the arms and body in the required series of actions, rather than having to program each one.
- **Care-bots**—the use of robots to supplement or replace the human care of sick or vulnerable people is a controversial one, but they can have some positive benefits. There are robots that help care for the elderly, either by providing 'company' (through Speech Recognition and NLU) or support through, for example, helping them remember things (Optimisation). Other AI-powered robots that are used in the medical field include the use of telepresence robots—these are mobile pods that can travel around hospitals but are connected via video and sound to human doctors at a remote location.
- **Service robots**—Some retailers are starting to use mobile robots to greet and serve customers. As with the hospital telepresence robots described above, these are mobile units with computer vision, speech recognition, NLU and optimisation capabilities built in. As well as being able to serve customers in a shop, different versions of the robots can take the role of waiters in restaurants or concierges in hotels. There are also interesting examples where robots are learning skills through reading material on the

internet—there is one case where a robot has learnt to make pancakes by reading articles on WikiHow.

- **Swarm robots**—these are a specific field of robotics where many small machines work collaboratively together. They rely heavily on the AI Optimisation capability, constantly evaluating the next best move for each of the swarm robots so that they can achieve the shared goal. Generally ground-based, but can be aerial or water-based, they are usually used in environments that are difficult for humans to work in, such as disaster rescue missions or, more controversially, in warfare (think autonomous armies). Autonomous vehicles will also be able to exploit swarm intelligence.

AI is also being used in human-like ways to help robots learn. Researchers have developed an approach where a humanoid robot learns to stand itself up by 'imagining' what it is like to stand up. Essentially it will run a series of simulations using a DNN, and then use a second system to analyse the feedback from the various sensors as it actually tries to stand up.

Cognitive Robotics, therefore, can be considered the physical embodiment of AI. Using input data from many different types of sensors the robot uses a combination of the Speech Recognition, Image Recognition, NLU and Optimisation capabilities to determine the most appropriate response or action. And, because it is AI, the system can self-learn, becoming more effective the more it is used.

Of course, physical cognitive robots also stir in us the fear that they will eventually be able to 'take over'. Software-based AI is not going to be a risk to the human race if it is stuck on a server; we can always pull the plug out. But physical robots could potentially have the ability to overpower us, especially if they are able to build better versions of themselves. There are already examples of cognitive robots being able to do this, but I refer you back to my earlier comments in the section on AI Understanding about the distinction between Artificial Narrow Intelligence and Artificial General Intelligence. There is, of course, an existential risk there; it's just that it won't need to concern us for a very long time.

AI and the IoT

The Internet of Things, or IoT, refers to simple physical devices that are connected to the internet. IoT devices include webcams, smart lights, smart thermostats, wearables and environmental sensors. Many people believe that IoT is, along with AI, one of the major technology trends of the decade.

There are billions of IoT devices in the world today, each one generating data or reacting to data, effectively creating and consuming big data on a grand scale, which is why AI has such a symbiotic relationship with the IoT.

IoT devices are being used in business today to

- manage preventative maintenance programs by analysing data from sensors embedded into the assets (e.g. escalators, elevators and lighting)
- manage supply chains by monitoring movement of products
- conserve energy and water usage of machines by analysing and predicting demand (smart meters, which monitor energy usage every 15 minutes, are already common in many homes)
- improve customer experience by providing personalised content based on IoT data
- increase crop yields by analysing field sensors to provide precise feeding and watering programs
- alleviate parking problems by matching empty spaces with cars and their drivers
- keep us fitter by monitoring and analysing our steps and exercise patterns through wearable devices

One of the headline successes of 'IoT plus AI' was when Google announced that they had been able to reduce the energy used for cooling in one of their data centres by 40%. By using DeepMind's AI technology, they were able to analyse data from many different types of sensors (such as thermometers, server fan speeds, and even whether windows were open or not), precisely predicting demand so that they could instruct the various machines to work at optimum levels.

Another area where IoT and AI are key enablers is in the development of Smart Cities. IoT devices are used to track and extract data for transport, waste management, law enforcement and energy usage. This data is analysed and turned into useful information by AI Clustering, Optimisation and Prediction systems, and then made available to the relevant authorities, the city's citizens and other machines. For example, sensors in street lights and urban furniture are able to measure footfall, noise levels and air pollution— this data is then used to prioritise the delivery of other services.

The biggest challenge to the proliferation of the IoT is the poor security that has been associated with it. Many devices do not have the ability to change passwords, and have only basic authentication. There have been recent cases of baby monitors allowing strangers to monitor their camera feed, internet-connected cars' entertainment systems and central locking systems being taken over by hackers, and, probably most worrying of all, medical devices being compromised so that a hacker could send fatal doses of medicines to drug infusion pumps. The good news is that these high-profile cases has meant that IoT security is very high on many technology company's agendas and there is now plenty of activity to fix all the issues. But if you are thinking of implementing an IoT strategy, do keep security at the top of your considerations.

As more and more IoT devices are used (it is estimated that there will be 50 billion devices by 2020), there will be more need for AI to make sense of all the data that is generated by them. The analysis that can be carried out can then be actioned by other IoT devices such as actuators and lights. As in the Smart Cities example, the real value will be the sharing and collaboration of this data across organisations, people and other machines.

AI and Crowd Sourcing

Sometimes AI simply isn't up to the job. Sometimes you will need to pull humans into the loop to help complete the process.

A classic example of when this sort of situation can arise is when the source documents for an automated process are handwritten. We already know that AI is very capable of extracting structured data out of unstructured documents but this really applies only when the unstructured documents are in an electronic format in the first place. Optical Character Recognition is able to take typed documents such as PDFs and convert them into an electronic format, but if the source is handwritten the challenge becomes a whole lot more difficult.

One solution to this is to use a Crowd Sourcing service. Crowd sourcing is where large numbers of people are engaged to carry out small parts of a process ('micro-tasks'). The engagement model is usually via the Internet, with each person paid a specific rate for every time they complete one of the micro-tasks.

In the example above, one document can be split into many different tasks so that one person is sent the First Name to read, another is sent the Last Name and a third the Social Security Number. Each person will look at the image of the handwriting and respond with the text that it represents. Because each person only sees a small part of the information, confidentiality is maintained. To increase accuracy, a single image can be sent to a number of people so that only the most common answer is selected. Specialised software is used to manage the interface between the customer and the crowd, including the splitting of the documents into smaller parts.

A second role for crowd sourcing in the AI space is to provide additional capability when the AI is not confident enough of its answers. This might be because the problem is too complex, or it hasn't seen that particular problem before (Fig. 4.1).

In this case, the AI will send the problem to a person to answer. For example, some AI systems are used to moderate offensive images on social media sites. If the AI is unsure whether a particular image is offensive or not, it will ask a human for their opinion. This approach is usually called 'Human In The Loop', or HITL.

An enhanced version of HITL is where the input from the human is used to actively train the AI system. So, in the offensive image example, the human's choice (offensive or inoffensive) would be sent back to the AI so that it can improve its learning and perform better in the future (Fig. 4.2).

A third use of crowd sourcing with AI is in the development of the training data. As I discussed in the Technology Overview section of Chap. 1, supervised learning approaches require data sets that are appropriately tagged (dog pictures tagged as dogs, cat pictures tagged as cats, etc.). Because of the large data sets that are required, the job of tagging all these data points is very laborious. Crowd sourcing can be used to satisfy that requirement. Google itself pays for tens of millions of man-hours collecting and labelling data that they feed into their AI algorithms (Fig. 4.3).

The most popular general crowd sourcing site is Mechanical Turk (owned by Amazon) but others, such as Crowd Flower, that focus on supporting the AI community, are also available. Some also have an Impact Sourcing angle as

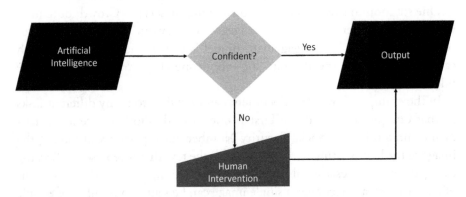

Fig. 4.1 Human in the loop

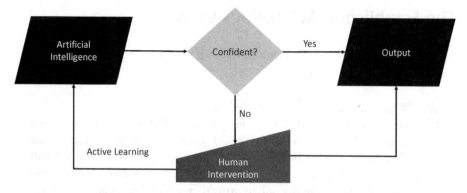

Fig. 4.2 Training through a human in the loop

Fig. 4.3 Crowd-sourced data training

well, where the people doing the work are disadvantaged in some way and this work gives them new opportunities.

So, despite the amazing things that AI can do, at some point it will probably still rely on people to do part of the job, whether that is training it or helping it make decisions.

The Established AI Vendor's View

This is an extract from an interview with Andrew Anderson, CEO of Celaton, a leading AI software vendor.

AB   What got you into the world of AI in the first place?

AA   The story began in 2002. I sold my previous software company to a much larger company that pioneered the delivery of software as a service, at that time they were described as an 'application service provider'. As the VP of product development, we went on to acquire, license and develop lots of software applications that solved many different challenges faced by organisations, all delivered as a service. However, despite access to all this technology we still couldn't solve the challenge of dealing with unstructured, varied (human created) data that flowed into organisations every day. It remained a manual, labour-intensive task.

I saw an opportunity to create a solution to the challenge and so in 2004 I bought my company back, along with my development team and set about trying to build an 'as a service' platform that could automate the processing of unstructured, varied data. My optimistic view then hoped that we would have a solution built within a year. The reality was that it took over six years to create a platform that had the ability to learn, and therefore enable us to describe what we had created as 'artificially intelligent'. In fact, it wasn't until we demonstrated it to analysts that they told us what we had created.

We spent a few years (2011–2013) trying different approaches until we realised that the greatest value we deliver was to large, ambitious organisations who deal with demanding consumers.

AB   Why do you think AI is being so talked about right now?

AA   AI is being talked about more than ever because it promises so much. Whether those promises are realised is yet to be seen. It's like a wonder drug that has huge potential but the enthusiasm needs to be tempered with a little reality.

What is emerging now are specialist AIs that are good at specific tasks rather than general-purpose AIs that are good at everything. Whilst there are many case studies where specific or narrow AI is delivering real benefits this is being mixed up with noise about general AI.

There's sanity in numbers and it takes time and effort to educate and convince enough organisations that they should try something but each new client means a new case study and each case study leads to new customers. These are the realities behind the hype, but the media often prefer to hear the more exciting stories of potential rather than reality.

Automation has been around since the wheel was invented and the world has continued to invent and innovate and that's why we see this continuous improvement in technology. The increased power of technology enables it to achieve outcomes that were previously out of reach.

Automation of any kind is also accelerating and with the emerging of AI it must seem that everything can be automated. What was previously considered the domain of human effort is now being eroded and this creates negative media.

In summary, I think it's being talked about more for its potential than its reality, although the reality (i.e. the case studies) are helping to fuel the future potential because people can actually see the future emerging.

AB   What value do your customers get out of your software?

AA   There's a different answer depending on the customer challenge. By understanding what customers are saying, then reacting and responding to them enables our customers to deliver better service faster, with fewer people. In summary, it enables them to achieve competitive advantage, compliance and improve their financial performance.

AB   What do customers or potential customers need to focus on if they are going to get the maximum value from AI?

AA   I think the key is not to focus on the technology but to understand the problem you want to solve. Many people (within organisations) have trouble understanding their problem and therefore they struggle to identify the technology that might help to solve it.

This is where experienced consultancy is important. Understand the problem, select and apply the most appropriate technology and then share the story.

AI is often considered to be the solution to all problems like a magic drug. The reality is there is no magic drug but there are lots of drugs that solve different problems. The key is to talk with someone who can help to understand the problem and prescribe the right 'drug'.

AB   Do you think the current hype is sustainable?

AA   I don't think that we are witnessing anything different with this hype. The industry does tend to get ahead of itself because hype serves it well, creating awareness that helps to capture the interest of customers. There's sanity in numbers.

The difference this time is that technology innovation is moving quicker and it's not so long after the thought has been considered that the invention can be demonstrated.

Regardless of how fast things move, it's the human that is the limiting factor. The bigger the problem that it solves, the more likely it is to be adopted by humans and therefore be successful.

AB   How do you see the market developing over the next 5 years or so?

AA   Hype will lead to reality, but by solving problems that really exist. There are a couple of areas that are particularly relevant:

- Natural interfaces. We're seeing this already with the likes of Amazon Echo and IoT appliances. There is some way to go but they will become more natural and to such an extent that humans will have to think less about how they communicate with technology.
- Broader AI. We're currently seeing specialist AIs that are good at doing specific things – these technologies will consolidate into broader solutions that are able to do lots of things well because they're made up of lots of vertical AIs.

And I particularly see some of the greatest and most profound use of AI in medicine, energy and, subject to world events, in the military too.

5

AI in Action

Introduction

Up to this point in the book I have been pretty theoretical, explaining the capabilities of AI and the types of technologies it requires to work. In this chapter I change focus to look at how AI is actually being used in businesses—real use cases of AI adding value and changing the way companies do business.

I have split the chapter into different themes which look at specific aspects of business and the ways that AI adds value: enhancing customer service, optimising processes and generating insights. There are overlaps between each of these areas (there are, for example, insights you can generate about your customers) but it gives a general, useful framework to describe how AI is being used in business.

In the chapter following this I describe the different types of benefits that should be considered in an AI Business Case. These can be broadly mapped against the themes:

- Enhancing customer service leads to revenue generation and customer satisfaction
- Optimising processes leads to cost reduction, cost avoidance and compliance
- Generating insights leads to risk mitigation, loss mitigation and revenue leakage mitigation

Each of these themes will also draw upon different capabilities from the AI Framework. All will use at least one of these capabilities, whilst others will exploit a few.

© The Author(s) 2018
A. Burgess, *The Executive Guide to Artificial Intelligence*,
https://doi.org/10.1007/978-3-319-63820-1_5

I have tried to take examples from a wide range of industries, and have deliberately not grouped the case studies into specific sectors. This is so you are not tempted to skip over sectors that are not immediately relevant to you—I am a firm believer that very different industry sectors can still learn from each other. Even if you work in the retail sector, for example, you may still get a spark of an idea from something in the utilities sector.

How AI Is Enhancing Customer Service

The 'front office' is one of the more active areas when it comes to implementing AI. This is because there is generally plenty of customer data available to work off, but it is also down to the proliferation of chatbots.

Chatbots come in all shapes and sizes, which is a rather polite way of saying that there are really good chatbots but also very bad ones. Chatbots, which aim to have a natural conversation with a customer through a typed interface, use NLU as their key AI capability. This means that, in theory, a customer who wants to change their address on the system (this is their 'intent' in AI-speak) can ask the chatbot in any way that they want ('I'm moving to a new house', 'I've got a new address', 'I'm not living at the same place any more', 'My postcode has changed', etc.) and it will still understand the underlying intent.

In reality, this is a big challenge for NLU, and most chatbots use a 'dictionary' of equivalent phrases to refer to (e.g. if the input is 'I've got a new address' or 'I'm moving to a new house', then the intent is 'Customer wants to change their address'). This, of course adds to the complexity of designing the chatbot, as every alternative phrase has to be identified and entered.

Some chatbots make extensive use of multiple-choice questions. So, instead of relying on understanding what the customer has typed, the chatbot will ask a question with a limited number of answers (e.g. Yes/No, Ask for a Balance/Make a Payment/Change Your Address, etc.). This can be more efficient and accurate but doesn't make for a natural conversation.

Another challenge for chatbots is how they move through the conversation. The simplest and most common approach is a decision tree, where each question will branch off to a new question depending on the answer. If the process being replicated is complex, this can lead to an over-complicated and burdensome decision tree. A better approach here is to have a cognitive reasoning engine do all of the 'thinking' whilst the chatbot gets on with the conversation. This provides much more flexibility in how the chatbot deals with the flow of the conversation. Cognitive reasoning systems were described in the Optimisation section of Chap. 3.

The purest AI approach to chatbots is to train them on thousands of human-to-human chat conversations, where each interaction has been tagged with the intent and whether it was productive or not. These chatbot systems effectively learn the knowledge map (a.k.a. ontology) through those historical interactions. The challenge with these systems is that there needs to be plenty of training data available, and they tend to very expensive to implement.

With chatbots, it is very much 'horses for courses'. A simple, 'free' chatbot system will be fine for very simple and non-critical interactions with customers, but because of the reputational risk involved, implementing chatbots is worth doing only if you do it right. The better the chatbot system, the better it will cope with the challenge.

In 2017 **Royal Bank of Scotland** introduced a chatbot to help answer a limited number of questions from customers. The chatbot, called 'Luvo', uses IBM Watson's Conversation capability to interact with customers who are using the bank's website or app. It was trialled for nearly a year with internal staff who manage relationships with SMEs before being released to a small number of external customers. At the time of writing, it is able to answer just ten defined questions, such as 'I've lost my bank card', 'I've locked my PIN' and 'I'd like to order a card-reader'.

Luvo demonstrates the cautious approach that enterprises are taking to implementing chatbots. Over time though, as the system learns from each interaction it has, the bank will allow it to handle more complex questions, build in greater personalisation and use predictive analytics to identify issues for customers and then to recommend the most appropriate action. The key objective is to free up more time for the human customer service agents to handle the trickier questions that customers may have.

In a similar two-step implementation, **SEB**, one of Sweden's largest banks, deployed IPsoft's Amelia software first in their internal IT Service Desk, and then to their one million customers. During the first three weeks of the trial, the chatbot-cum-avatar helped to answer the staff's IT issues, resolving around 50% of the 4000 queries without human interaction. During 2017, it was rolled out to help manage interactions with the bank's customers (and named 'Aida'). Three processes were initially chosen as candidates: providing information on how to become a customer, ordering an electronic ID and explaining how to do cross-border payments.

Although Amelia learns from historical interactions and uses sentiment analysis, it is also able to follow defined workflow paths to ensure compliance with banking regulations. Whereas most chatbots can be described as probabilistic, providing likely results, Amelia can be considered a deterministic system because, once it has worked out the intent, it can, where possible, carry out the required actions on the enterprise systems (in a similar way that RPA tools would). IPsoft is a US-based company, and this was their first non-English deployment of their solution.

From an infrastructure point of view, SEB decided to install the Amelia technology on their own servers rather than as a cloud solution. This was due to concerns over the compliance and legal concerns that a cloud deployment raises. For IPsoft's financial services customers, this is the most common approach.

The deployment of Amelia is very much aligned with the bank's overall strategy, which includes the line "SEB will focus on providing a leading customer experience, investing in digital interfaces and automated processes".

The US flower delivery service **1-800-Flowers** deployed a simple chatbot that allowed customers to place orders through Facebook Messenger. It is a linear, decision tree system with an NLU chat interface that took around three months to develop and test. This makes it rather constrained in what it can do, but additional functionality and complexity is being rolled out. Following the initial two months of operation, 70% of the orders being placed through the chatbot were from new customers, and were dominated by 'millennials' who tend to already be heavy users of Facebook Messenger.

As well as taking orders, the chatbot can direct a customer to a human agent, of which there may be up to 3,500 at any one time. They have also implemented an integration with Amazon Alexa, and a concierge service that is powered by IBM Watson. Together, these digital initiatives have attracted 'tens of thousands' of new customers to the brand. They also provide up-to-the-minute behaviour data for the company which can then influence real-time marketing activity such as promotions.

Another company that has used IBM Watson to enhance customer engagement is **Staples**, the stationery retailer. They have implemented a range of different ways for people to buy their products as easily as possible, including email, Slack, mobile app and (honestly) a big red button. The button is similar to Amazon's Alexa in that it can understand voice commands (although it does need a physical press to activate and deactivate it). The mobile app can also understand voice commands as well as being able to identify products from photographs taken. All these channels make the buying process as frictionless as possible for the customer, and therefore has a direct and positive impact on revenue for the retailer.

As well as chatbots, recommendation engines are another common AI technology that is used to enhance customer service (and, of course, drive revenue). **Amazon** and **Netflix** have the most well-known recommendation engines, and these are deeply embedded in the normal workflow of how the customer engages with the companies. All the required data—the individual customer's buying and browsing behaviours and the historical data of all the other customers' behaviours—are available through those normal interactions; that is, there is no additional work that the customer has to do to enrich the data set.

In some cases, the recommendation engines will require additional information from the customer and/or the business in order to work effectively. **North Face**, a clothing retailer, has implemented a recommendation engine for their customers who want to buy jackets. Based on IBM Watson this solution, called XPS, uses a chatbot interface to ask a series of refining questions so that it can match the customers' requirements with the product line. According to North Face, 60% of the users clicked through to the recommended product.

Another clothing retailer, **Stitch Fix**, uses a slightly different approach by deliberately including humans in the loop. Its business model involves recommending new clothes to their customers based on a selection of information and data that the customer provides (measurements, style survey results, Pinterest boards, etc.). All this structured and unstructured data is digested, interpreted and collated by the AI solution, which sends the summary, plus anything that is more nuanced (such as free-form notes written by the customer) to one of the company's 2,800 work-from-home specialist human agents, who then select five pieces of clothing for the customer to try.

This is a good example of where the AI is augmenting the skills and experience of the human staff, making them better at their jobs as well as being more efficient. Having humans in the loop (HITL is the acronym for this) also makes experimenting easier, as any errors can quickly be corrected by the staff. To test for bias, the system varies the amount and type of data that it shows to a stylist—it can then determine how much influence a particular feature, say a picture of the customer or their address, can make on the stylist's decisions. On top of all this, the data that they gather about all of their customers can also be used to predict (and influence?) general fashion trends.

Other customer service AI solutions don't rely on chatbots or recommendation engines to provide benefits. **Clydesdale and Yorkshire Banking Group** (CYBG) is a medium-size bank in the United Kingdom, having to compete with the 'Big Four' of Barclays, HSBC, Lloyds and RBS. Their digital strategy includes a new current account, savings account and app package, called 'B'. It uses AI to help manage the customer's money: it allows you to open account, and, once opened, will learn the patterns of usage so that it can predict if you might run out of funds in your accounts, and suggest ways to avoid unnecessary bank charges. The bank claims that an account can be opened in 11 minutes. Clearly, customers avoiding bank charges will result in lower revenue, but the hope is that it will attract enough new customers that this revenue loss will easily be outweighed by the benefits of having the additional funds.

Virgin Trains 'delay/repay' process has been automated through the application of AI. After implementing Celaton's inSTREAM AI software to categorise inbound emails (see next section) the train operating company used the same software to provide a human-free interface for customers to automatically claim refunds for delayed trains.

The Asian life insurance provider **AIA** has implemented a wide range of AI initiatives that all impact how they engage with their customers, including insight generation from prospective customers, enhanced assessment of customer needs, 24×7 chatbots for online enquiries, inbound call handling by NLU-based systems, enhanced compliance of sales, personalised pricing, dynamic underwriting and an augmented advice and recommendation engine.

Some companies are building AI into the core of their customer applications. **Under Armour**, a sports clothing company that has a portfolio of fitness apps, uses AI in one of those apps to provide training programs and recommendations to the users. The AI takes data from a variety of sources, including the users' other apps, nutritional databases, physiological data, behavioural data and results from other users with similar profiles and objectives. It then provides personalised nutrition and training advice that also takes into account the time of day and the weather.

Other examples of where AI can enhance customer service are: optimising the pricing of time-critical products and services such as event ticketing; optimising the scheduling of real-time services such as delivery and departure times; creating personalised loyalty programs, promotional offers and financial products; identifying candidate patients for drug trials and predicting health prognoses and recommending treatments for patients.

There is clearly a balance between using AI to serve customers better, and using AI to gather information about customers. In some of the cases I have described, the same AI capability will do both. As I have reiterated a number of times, for anyone looking to implement a 'front office' AI solution, it is important to first understand your objectives. The danger of trying to ask your AI to do too much is that you compromise on one objective or the other. I discuss approaches and examples of how to generate customer insights later in this chapter.

How AI Is Optimising Processes

If the previous section was all about enhancing the front office, then this section focuses on what happens behind the scenes in the back office operations. The benefits here are centred around, although not exclusively, reducing cost (directly or through avoidance) and improving compliance.

One of the key capabilities exploited in the back office is being able to transform unstructured data into structured data, that is, from the Image Recognition, Voice Recognition and Search capabilities.

Tesco have implemented a number of AI-powered solutions to improve the productivity of their stores. They are using image recognition systems to identify empty shelves on stores, called 'gap scanning'. They are experimenting with physical robots that travel down the aisles during quiet periods to film the shelves so that they can measure stock availability and inform staff to replenish them where necessary. Not only does this save the time of the staff (who would normally have to use a diagram on a piece of card as the template) but it also reduces revenue loss due to stock unavailability.

For their home delivery service Tesco have implemented optimisation systems that minimise the distance the picker walks around the store collecting the goods, and also a system that maximises the productivity of the delivery vans through effective route planning and scheduling. Interestingly, in a meta-example, they have used AI to help place the many products into the necessary taxonomy that AI systems use for classification.

AIA, the Asian insurance provider I mentioned in the previous section, have also implemented initiatives in their servicing and claim functions, including service augmentation, investment management, medical second opinions, risk management, fraud management, claims adjudication and health and wellness advice to their customers. AIA have been very clear that transforming their business requires end-to-end transformation with AI at the core.

Insurance claims processing is a prime candidate for enhancement through AI and RPA. **Davies Group**, a UK-based insurance firm that provides a claims processing service to their own insurance customers, has automated the input of unstructured and semi-structured data (incoming claims, correspondence, complaints, underwriters' reports, cheques and all other documents relating to insurance claims) so that it is directed into the right systems and queues. Using Celaton's inSTREAM solution, a team of four people process around 3,000 claims documents per day, 25% of which are paper. The tool processes the scanned and electronic documents automatically, identifies claim information and other meta-data, and pastes the output into databases and document stores ready for processing by the claims handlers and systems (which could, of course, be either humans or software robots). It also adds service meta-data so the performance of the process can be measured end to end. Some documents can be processed without any human intervention, and others need a glance from the human team to validate the AI's decisions or fill in missing details.

AI can also be used in the claims process to identify fraud (by looking for anomalous behaviours), to help agents decide whether to accept a claim or not (using cognitive reasoning engines) and by using image recognition to assess damage. **Ageas**, a Belgian non-life insurance firm, has deployed Tractable, an AI image recognition software, to help it assess motor claims. By scanning images of damage to cars, the software is able to assess the level of damage, from repairable to being a write-off. Through the process, it also helps identify fraudulent claims.

Image recognition has also been used in other industry sectors. **Axon**, which used to be known as Taser, the manufacturer of police equipment, are now using AI to tag the thousands of hours of video that is recorded from the body cameras that they manufacture. The video recordings were previously available to police forces on subscription but, due to the sheer volume of recordings, they were rarely used as evidence. Now that the film is being tagged, the crime agencies will be able to automatically redact faces to protect privacy, extract relevant information, identify objects and detect emotions on people's faces. The time saved, from having to manually write reports and tag videos, will be huge, but the ability to make better use of the available evidence will likely have a bigger impact. For Axon, it will encourage more police forces to buy their body cams.

Data from video feeds has been used as the core of the business model of **Nexar**, an Israeli-based technology company. They are giving away for free their dash cam app which helps drivers anticipate accidents and problems in the road ahead. By using the data provided by the apps from their user base (which is mandatory if you want to use the app) the system predicts the best course of action, for example to brake hard if there is an accident up ahead. This model relies heavily on having large amounts of data available, hence the up-front giveaway. Rather controversially, the company can use the data for 'any purpose', including selling it to insurance companies and giving it to the government. Whether this is a sustainable business model remains to be seen, but it does demonstrate how business models are transforming to focus more on the data than the service being provided to the customers.

A particularly worthwhile use of image recognition is in the medical field. Probably the most widely reported has been the use of IBM's Watson platform to analyse medical images, such as X-rays and MRI scans, so that it can help identify and diagnose cancer cells. The real benefit of this system is that it can cross-reference what it sees with the patient's (unstructured) medical records and thousands of other images that it has learnt from. This approach is already spotting characteristics of tumours that medical staff hadn't seen—they are finding that peripheral aspects of the tumour (e.g. those distributed around, rather than in, the tumour) have a much greater influence on predicting malignancy than at first thought. These systems are currently working in two hospitals: **UC San Diego Health** and **Baptist Health South Florida**. (See the section on Generating Insights for more examples of how AI is helping fight cancer.)

Deutsche Bank are using the speech recognition technology to listen to recordings of their staff dealing with clients, in order to improve efficiency, but, more importantly, ensure compliance to regulations. These conversations are transcribed by the AI and can be searched for specific contextual terms. This used to be the job of the bank's auditors, and would require listening to hours of taped recordings, but the AI system can do the same thing in minutes. The bank is also using other AI capabilities to help it identify potential customers based on the large amount of public information that is available on people.

The legal sector is currently going through a big transition period as it starts to adopt new technologies, including AI. Inherently risk averse, and never an industry to change quickly, the legal sector is starting to feel the benefit of AI particularly through implementing Search and NLU technologies to help make sense of contracts.

Legal contracts can be described as semi-structured documents—although they follow some general rules (they usually include parties to the contract, start and end dates, a value, termination clauses, etc.), they are verbose and variable. There are now a number of software solutions on the market which, together, work across the full contract lifecycle—some are able to search document repositories to identify where contracts are; others can 'read through' the contracts, identify specific clauses and extract the relevant meta-data (such as the termination date and the limits of liability). Some have the capability to compare the real contracts with a model precedent and show all instances where there is a material difference between the two. All of this allows large enterprises and law firms to manage their contracts much more effectively, especially from a risk point of view.

McKesson, a $179 billion healthcare services and information technology organisation, uses Seal's Discovery and Analytics platform to identify all of their procurement and sourcing contracts across the business (they employ 70,000 people) and store the documents in a specific contract repository. When in the repository, the contracts could be searched easily and quickly, saving some staff hours per day. The meta-data from the contract analysis allowed them to identify potential obligation risks as well as revenue opportunities and savings from otherwise hidden unfavourable payment terms and renewal terms.

A Magic Circle law firm, **Slaughter & May**, adopted Luminance AI software to help manage the hundreds of Merger and Acquisition (M&A) transactions they do a year. This area is a particular challenge because of the complexity involved (thousands of documents, many jurisdictions) and the intensity of the work—the law firm was concerned that some of the junior lawyers who were tasked with managing the M&A 'data room' (the contract repository created specifically for the deal) would eventually burn out. Luminance is used to cluster, sort and rank all of the documents in the data room, with each document assigned an anomaly score to show how it differs from the ideal model. Whereas in normal due diligence exercises only around 10% of the documents are analysed, due to the sheer volume, Luminance is able to analyse everything. It takes about an hour for it to process 34,000 documents. Overall, it halves the time taken for the whole document review process.

One of the benefits of having the AI systems carry out 'knowledge based' works such as this is that it allows lawyers to focus on complex, higher-value work. **Pinsent Masons**, a London-based law firm, designed and built in-house a system, called TermFrame, which emulates the legal decision-making workflow. The system provides guidance to lawyers as they work through different types of matters, and connects them to relevant templates, documents and precedents at the right time. Taking away much of the lower-level thinking that this requires means that the lawyers can spend more time on the value-add tasks.

As well as calling on Search and NLU capabilities, companies are also using NLG solutions to optimise some of their processes. Using Arria's NLG solution, the **UK Meteorological Office** can provide narratives that explain a weather system and its development, with each one configured with different audiences in mind. The **Associated Press** (AP), a news organisation, uses Wordsmith to create publishable stories, written in the AP in-house style, from companies' earnings data. AP can produce 3,700 quarterly earnings stories which represents a 12-fold increase over its manual efforts.

Artificial intelligence can also be used to enhance the IT department as well ('Physician, heal thyself'). There are a number of AI capabilities that can be deployed, both in the support function and for infrastructure management. Generally, AI brings a number of important aspects to the IT function: the AI systems can be trained from run-books and other sources; they will continue to learn from 'watching' human engineers; they can proactively monitor the environment and the current state; they can identify trends in system failures and they are able to cope with the inherent variability of IT issues.

We have already seen how chatbots can support customer engagement, but they can also be used with staff as well, particularly where there are a large number of employees, or a large number of queries. Although chatbots can be deployed on any type of service desk, including HR, they are most commonly found in IT.

As in the **SEB** example discussed in the previous section, AI Service Desk agents are able to receive customer enquiries and provide answers by understanding what a customer is looking for and asking the necessary questions to clarify the issue. The more advanced systems will, if they cannot help the customer themselves, raise the issue to a human agent, learning how to solve the problem itself for future situations.

AI can also be used to manage a company's IT infrastructure environment. The AI systems tend to work as meta-managers, sitting above and connecting all the various monitoring systems for the networks, servers, switches and so on. They include both proactive and predictive monitoring (using AI) and execution of the necessary fixes (using software robots). Key performance indicators such as downtime can be significantly improved using these systems. And, by tackling many of the mundane tasks, the systems free up the IT engineers to focus on higher-value, innovative work.

TeliaSonera, a European telecoms operator, implemented IPCentre to help manage its infrastructure of 20 million managed objects, including 12,000 servers, and have subsequently seen cost savings of 30%. A **New Yorkbased investment firm** used the same solution to help it fix failed fixedincome securities trades due to system issues. Eighty per cent of the failed trades are now fixed without human intervention, and the average resolution and fix time has been reduced by 93% (from 47 minutes to 4 minutes). This has resulted in a staff reduction of 35%.

Google, a company with many very large data centres, turned to its AI subsidiary, DeepMind, to try and reduce the cost of powering these facilities. The AI worked out the most efficient methods of cooling by analysing data from sensors amongst the server racks, including information on aspects such as temperatures and pump speeds. From the application of the AI algorithms, Google was able to reduce cooling energy demand by 40%, and to reduce the overall energy cost of the initial data centre by 15%. The system now controls around 120 variables in the data centres including fans, cooling systems as well as windows.

Another example of where AI can optimise processes includes the ability to model real-world scenarios on computers, with the most obvious example being the prediction of long-term weather conditions. But this modelling approach has been used to test physical robotic systems so that the designers can make adjustments to their robots virtually without the fear of them falling over and breaking. **Facebook** have created a specific version of Minecraft, a virtual-world game, for this sort of purpose. AI designers can also use the environment to help teach their AI algorithms how to navigate and interact with other agents.

In this section, we have seen examples of how AI can optimise the back office processes of retailers, banks, insurance firms, law firms and telecoms companies, across a wide range of functions, including claims management, compliance and IT. Other use cases for AI in optimising processes include: optimising logistics and distribution of inventory in warehouses and stores; real-time allocation of resources to manufacturing processes; real-time routing of aircraft, trucks and so on; optimising the blend and timing of raw materials in refining and optimising labour staffing and resource allocation in order to reduce bottlenecks.

How AI Is Generating Insights

In the previous two sections I looked at how AI can enhance customer service and optimise processes. But, in my mind, the biggest benefits from AI are when it can deliver *insight*. This is where new sources of value are created from the data that already exists, enabling better, more consistent and faster decisions to be made. AI can therefore enable a company to mitigate risks, reduce unnecessary losses and minimise leakage of revenue.

One of the most effective uses of AI to date has been in the identification of fraudulent activity in financial services. The benefit here is that there is plenty of data to work from, especially in retail banking. **PayPal** process $235 billion in payments a year from four billion transactions by its more than 170 million customers. They monitor the customer transactions in real time (i.e. less than a second) to identify any potentially fraudulent patterns—these 'features' (in AI-speak) have been identified from known patterns of previous fraudulent transactions. According to PayPal, their fraud rate is just 0.32%, compared to an industry average of 1.32%.

The American insurance provider **USAA** is similarly using AI to spot identity theft from its customers. It is able to identify patterns in behaviour that don't match the norm, even if it is happening for the first time. **Axa**, another insurance firm, has used Google's TensorFlow to predict which of its customers are likely to cause a 'large-loss' car accident (one that requires a pay-out greater than $1,000). Normally up to 10% of Axa's customers cause a car accident every year, but about 1% are 'large-loss'. Knowing which customers are more likely to be in that 1% means that the company can optimise the pricing of those policies.

The Axa R&D team initially tried a Decision Tree approach (called Random Forest) to model the data but were able to achieve only 40% prediction accuracy. Applying deep learning methods on 70 different input variables, they were able to increase this to 78%. At the time of writing, this project is still in its proof-of-concept (PoC) phase but Axa are hoping to expand the scope to include real-time pricing at point of sale and to make the system more transparent to scrutiny.

Another company that is using AI to optimise pricing in real time is **Rue La La**, an online fashion sample sales company who offer extremely limited-time discounts on designer clothes and accessories. They have used machine learning to model historical lost sales in order to determine pricing and predict demand for products that it has never sold before. Interestingly, from the analysis they discovered that sales do not decrease when prices are increased for medium and high price point products. They estimated that there was a revenue increase of the test group of almost 10%.

Otto, a German online retailer, is using AI to minimise the number of returns they get, which can cost the firm millions of euros a year. Their particular challenge was the fact that they knew that if they got the orders to their customers within two days then they would be less likely to return them (because it would be less probable for them to see the same product in another store at a lower price). But, they also knew that their customers liked to get all their purchases in one shipment, and, because Otto sources the clothes from other brands this was not always easy to achieve.

Using a deep learning algorithm, Otto analysed around three billion historical transactions with 20 variables so that it could predict what customers would likely buy at least a week in advance. Otto claims that it can predict what will be sold within 30 days to an accuracy of 90%. This means it can automate much of the purchasing by empowering the system to auto-order around 200,000 items a month. Their surplus stock has reduced by a fifth and returns have gone down by more than two million items a year.

Goldman Sachs, the global investment bank, has implemented a range of automation technologies to both improve decision-making processes and also to reduce headcount. They started by automating some of the simpler trades that were done at the bank—in the year 2000 there were 600 equity traders working in the New York headquarters; at the start of 2017 there were just two. The average salary of an equity trader in the top 12 banks is $500,000, so the savings have been significant. The majority of the work is carried out by automated trading systems that are supported by 200 engineers (one third, or 9,000, of Goldman Sachs employees are computer engineers).

To automate more complex trades such as currency and credit, cleverer algorithms are required. Where Goldman Sachs has automated currency trading, they have found that one computer engineer can replace four traders. They are also looking at building a fully automated consumer lending platform that will consolidate credit card balances. This is an innovation that has been incubated internally at the bank's New York office by small 'bubble' teams. As Marty Chavez, the company's deputy chief financial officer, says: there is plenty of empty office space to use up.

The UK-headquartered bank **HSBC** has been running five proof-of-concepts (PoCs) that will utilise Google's AI capabilities. Because their 30 million or so customers are engaging online much more, their data store has increased from 56 petabytes in 2014 to more than 100 petabytes in 2017. This means they are now able to mine much more value and insight from it.

One of the PoCs is to detect possible money laundering activity. Just as with PayPal and USAA, they are looking for anomalous patterns that they can then investigate with the relevant agencies to track down the culprits. As well as improving detection rates, using the AI software (from Ayasdi) means that there are less false-positive cases that have to be investigated, which saves the time of expensive resources. In the case of HSBC, they managed to reduce the number of investigations by 20% without reducing the number of cases referred for further scrutiny. HSBC are also carrying out risk assessments using Monte Carlo simulations (these were described in the Optimisation section in Chap. 3). From these they will be better able to understand their trading positions and the associated risks.

Some of the more contentious uses of AI in business are in recruitment and policing. The main contention is how bias can be (or even perceived to be) built into the training data set, which is then propagated into the decision-making process. In Chap. 8, I will discuss this challenge and the potential mitigating approaches.

The **Police Force of Durham**, a county in the North East of England, have started using an AI system to predict whether suspects should be kept in custody or not (in case they might reoffend). In this case, there are lots of data available, from previous cases of reoffenders and non-reoffenders over five years, that can be used to predict future reoffenders. At the time of writing it would seem that there still a number of issues with this approach, the biggest of which is probably with the validity of the data—this comes only from the region's own records, so excludes any offending that the suspect may have done outside Durham. Despite the data challenges, the system accurately forecast that a suspect was a low risk for recidivism 98% of the time, and 88% for whether they were a high risk. This shows how the system deliberately errs on the side of caution so that it doesn't release suspects if its confidence is not high enough.

In recruitment, a number of firms are taking the plunge into using AI to help short-list candidates (although there are only a few who will currently admit to it). **Alexander Mann**, a Human Resource Outsourcing Provider, initially automated some manual tasks such as interview scheduling and authorising job offers. They have recently introduced some AI software (Joberate) to help them find candidates that are good two-way matches with the job requirement. The software analyses both the candidate's CV and publicly available social media feeds to create profiles of the candidates.

But AI can mean more than just mitigating business risks. Insights from big data analysis can also be used to help fight diseases, especially cancer. AI has been applied to the molecular analysis of gene mutations across the whole body—usually cancer treatment research relates to specific organs—which means that treatments that have been developed for, say, breast cancer could be used to treat colorectal cancer. Through this approach personalised treatments become possible—statistically significant meta-studies have looked at how much benefit there was in matching the molecular characteristics of the tumour of a patient with their treatment. This matching resulted in tumours shrinking by an average of 31% compared to 5% from a non-personalised approach.

AI is also being used to develop cancer drugs. A biotech firm, **Berg**, fed as much data as it could gather on the biochemistry of cells into a supercomputer so that the AI could suggest a way of switching a cancerous cell back to a healthy one. The results have so far been promising, leading to the development of a new drug, and, of course, those results have been fed back into the AI to further refine the model.

The quality of healthcare can also be monitored using AI. The **Care Quality Commission** (CQC), the organisation that oversees the quality of healthcare delivered across the United Kingdom, implemented a system to process large numbers of textual documents and understand the opinions and emotions expressed within them. The CQC is now able to manage the influx of reports with fewer people and, most importantly, can apply a consistent scoring methodology. The use of sentiment analysis (which was described in the NLU section of Chap. 3) is also used in business-to-consumer companies, such as **Farfetch**, a UK online retailer, to understand in near real time what their customers think of their products and services. The companies are then able to respond quickly and get a better understanding of their customers' needs.

Other use cases for AI in optimising processes include: predicting failure for jet engines; predicting the risk of churn for individual customers; predicting localised sales and demand trends; assessing the credit risk of loan applications; predicting farming yields (using data from IoT sensors) and predicting localised demand for energy.

From the above examples, it should be clear that AI has huge potential to unearth hidden value from a company's data, helping manage risks and make better, more informed, decisions. There are certainly challenges around the lack of transparency and unintended bias, but, managed correctly the data can provide insights that would have been impossible for a human to discover.

The Established AI User's View

This is an extract from an interview with John Sullivan, CIO of Virgin Trains West Coast, one of the major train operating companies in the United Kingdom.

AB Tell me how you first came across AI.

JS Well, I actually studied artificial intelligence at college. At that time, it was obvious to me how useful AI would be but it wasn't yet very practical, particularly in a business setting. But I did get a good understanding of what it could do and why it is different from traditional systems.

AB And when did you start looking for AI opportunities at Virgin Trains?

JS In my current role as CIO, I was interested in looking for applications for AI that could solve some real issues. We had a customer relationship challenge in that we were getting lots of queries by email that took a long time for people to deal with. Some of the same questions were coming up again and again, day in, day out.

So we looked at implementing Celaton's inSTREAM solution to improve the service we were giving to the customers. By using AI we were able to respond to the majority of these queries quicker and more consistently. It also made it more efficient for us as a business and the work became more interesting for the staff.

AB What does the AI software do?

JS Basically, it reads all of the incoming emails, which of course are all written in free-form, and works out what the customer wants – it understands whether their email is a question, a complaint or a compliment and sends it to the right agent within the organisation to deal with. It also does a lot of validation work at the same time – so if a customer is writing about a specific train, the system will check that that train actually ran, and will also see if there are any similar queries relating to that journey – that all helps us prioritise the email.

We've managed to reduce the amount of effort required at that initial stage by 85%, which far exceeded our expectations. All of the people that were doing that very mundane work are now doing much more interesting stuff dealing with the more challenging questions we get.

AB So, do humans still deal with the remaining 15%?

JS That's right. When we started off we tried to identify which types of queries the AI could respond to and which ones a human would need to do. Because the system learns as it goes along, more and more of the work can now be done by the AI. So the agents only work on the really tricky ones, as well as over-seeing the work done by the AI. This oversight task is important for Virgin as we want to get the tone of the responses just right so that they align with the brand. We call it 'bringing out the Richard Branson'.

AB   So apart from improving customer service, what other benefits have you seen?

JS   The really important thing for us is that it has allowed the business to be scalable. We now have a core team of human agents that is stable, so that any increase in volume can be handled by the AI. Email volumes are now not an issue for us.

AB   How did you approach the implementation?

JS   We found Celaton quite quickly and immediately got on well with them. They came in and we did a prototype with the Customer Relationship team. Doing 'small speedboat' projects always works best for me for trials. It's always better than trying to build an ocean liner, which we know takes too long, and are difficult to stop!

AB   What challenges were there in implementing the system?

JS   Change management is always going to be a challenge in projects like this. The CS team, though, got on board really quickly because they actually wanted it. Remember – this was a system that was going to take away all of those mundane, repetitive queries for them. They couldn't wait!

The other thing that can be tricky with an AI project is getting the data right. We definitely needed to involve the IT team, but, because we had a good vendor in, Celaton knew what the challenges would be so we could prepare as much as possible. We relied on their resources quite a bit, as we didn't really have any AI capability in-house at the time, but they were good people who could explain everything in a simple manner to our people—they weren't just a 'technical' supplier.

AB   Thanks John. And finally, what advice would you give anyone just starting out on their AI journey?

JS   Well, the communications, which I just mentioned, are a really important aspect. You need to be able to articulate what AI is and how it works – don't assume that even the CIO knows these things. If I were to do this whole thing again I'd probably bring somebody in that could focus on this. Almost like doing internal marketing for the project.

I also think it is vital to open your mind up to what AI can do—again external input can be useful here. We have Innovation Days to look at developments we are doing with the trains, and we really should try and do the same sort of thing for AI as well. It's all about trying to understand the art of the possible.

6

Starting an AI Journey

Introduction

I am often asked by executives, "I need AI in my business—how can I implement it?". This is, of course, the wrong question to ask. The much more apposite question is, "I have some big business objectives/challenges—how can AI help me deliver or address them?" Ideally, executives should be pulling AI in to where it is needed the most and will provide the greatest value. In reality, though, it is usually a bit of push and pull—AI is such a new subject and has such huge potential to disrupt whole business models that it would be foolish to consider it only in the context of your existing business strategy.

So, if businesses are already implementing AI and deriving real value from it, as we have seen in the case studies presented in this book, how did they actually start their journey? How did they discover, plan and implement the solutions? This chapter will cover the approach to creating a meaningful and workable AI strategy.

I have deliberately blurred the lines between an 'AI Strategy' and an 'Automation Strategy'. As is hopefully clear from Chap. 4 when I discussed the associated technologies, AI is rarely the only answer. Usually other automation technologies are required, such as RPA, cloud and IoT. But with this book's focus on AI, I will be looking at the automation strategy very much through the lens of AI, mentioning the other technologies only where I think it is necessary for clarity.

If I could summarise the best approach to maximising value from AI it would be: think first, try something out, then go large. It's very tempting to jump straight in and create a PoC or bring in some AI software, but creating

© The Author(s) 2018
A. Burgess, *The Executive Guide to Artificial Intelligence*,
https://doi.org/10.1007/978-3-319-63820-1_6

an automation or AI strategy, one that matches your ambitions, is based on your business strategy and has explored the business at a high level for the prime opportunities, will give you, by far, the best foundation to maximise value.

Once you have the AI strategy sorted, then you can start trying things out. This isn't mandatory, but it usually helps to build knowledge, trust and momentum within the business. Your initial steps could be through a pilot, a PoC, a hard prototype or a purchased piece of software. I will explain each of these and discuss their pros and cons in this chapter.

But once the initial efforts are in and proven, then you will need to look at pushing on quickly. There can be, at this stage, what I call the AI Bermuda Triangle—lots of really good ideas that were heading in the right direction mysteriously disappear from the radar. This happens with other technology programs as well—once the effort of getting that first piece of work off the ground has been done, everyone sits back and becomes distracted by other things, motivations dissipate and momentum is lost. Once this happens it is very difficult to get it back on track again. Therefore, the time to really push forward is as soon as that pilot or PoC or prototype is showing promise and demonstrating value.

And that plan really needs to be big and bold if it is to escape from the AI Bermuda Triangle. Creating an AI roadmap is a key part of the AI strategy—after all, how are you going to go on an AI journey without a map to guide you? But it's more than just a list of projects—it needs to describe how AI will be industrialised within the business, and I'll cover those aspects in more detail in the penultimate chapter.

So, armed with the knowledge of what AI is capable of and how other businesses are using it, it is now time to start your own AI journey.

Aligning with Business Strategy

In my work as a management consultant and AI specialist, I know that the one single activity that will ensure that maximum value is extracted from any AI program is to first create an automation strategy that aligns with the business strategy, but also one that challenges it.

To align the automation strategy with the business strategy it is necessary to understand the benefits and value that will be derived from that overall strategy. Business strategies, if they are written succinctly, will have a handful (at most) of strategic objectives—these might be things like 'Reduce the cost base', 'Reduce the exposure to internal risk' or 'Improve customer service CSAT score'.

Each of these strategic objectives would deliver benefits to the business: 'Reducing the cost base' would deliver lower costs through, for example, not hiring any new staff, not taking on any additional office space or minimising travel. 'Reducing the exposure to risk' might be delivered through reducing the number of unnecessary errors made by Customer Service staff, or improving reporting and compliance. And 'Improving Customer Service' could be fulfilled by improving the average handling time (AHT) of inbound queries, reducing the unnecessary errors (again) and enabling 24×7 servicing.

The important thing from our perspective in creating an automation strategy is to understand its role in enabling some or all of these benefits.

In a hypothetical example: AI Search could be used to reduce AHT by reading incoming documents; Optimisation tools could be deployed to provide knowledge support to the customer service agents; self-service could be enabled through a number of AI and RPA tools so as to reduce the demand on bringing new staff and renting more office space; Clustering and Search capabilities could be used to offer deep reporting insights into the business information; Search could be used to identify areas of non-compliance against a constantly updating regulatory database and chatbots (using NLU) could be used to provide a 24×7 service desk. RPA could also be brought in to eliminate errors. Each of these tools then becomes an enabler to the business strategy benefits (Fig. 6.1).

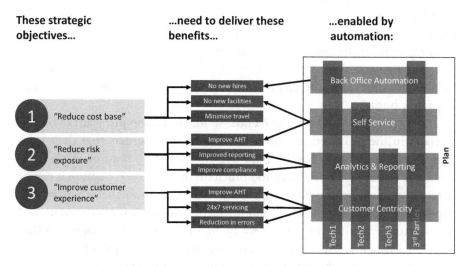

Fig. 6.1 Aligning with the business strategy

So, before we have even started creating the automation strategy, we must know what it should achieve. Then it is a case of working backwards to determine the capabilities, technologies, people, plan and so on that will deliver those objectives.

Of course, there is nothing to stop you from implementing AI simply because you want to implement AI. In fact, demonstrating to shareholders or customers that you are an innovative, forward-thinking company because you are using AI can be of great benefit, and much PR can be made of the fact. But, generally, aligning your Automation Strategy firmly with the Business Strategy will ensure that the benefits you are hoping to achieve will deliver long-term, sustainable value to the business.

Understanding Your AI Ambitions

The second major aspect to consider before starting your AI journey is to understand, as much as possible, where you want to end up. That may sound obvious, or you may think why should you bother thinking about that now when you haven't even started anything yet. But, as with any journey, it is crucially important to know your destination. In AI terms, you won't be able to know exactly where you will end up (it's a voyage of discovery, after all) but you should at least understand your initial AI ambitions.

These ambitions can range from anywhere between 'we just want to say we've got some AI' to creating a completely new business model from it. There are no right and wrong answers here, and your ambition may change along the way, but knowing your overall aspirations now will mean that you can certainly start on the right foot and be going in the right direction.

I would consider four degrees, or levels, of ambition, which I have called: ticking the AI box, improving processes, transforming the business and creating a new business.

The first of these, '**ticking the AI box**', is for those who just want to be able to claim in their marketing material and to their customers that their business, service or product has AI 'built in'. This approach is surprisingly common, but I'm not going to focus too much on this approach because it can be covered by selecting the most appropriate elements from the other types of approaches. To be honest, you could justifiably say that your business uses AI now because you filter your emails for spam, or that you use Google Translate occasionally. Many businesses that have deployed simple chatbots claim to be 'powered by AI', which is factually true but a little disingenuous.

The first serious step to extracting value from AI will be in **improving the processes** that you already have, without necessarily changing the way the function or the business works. AI is being used to make existing processes more efficient and/or faster and/or more accurate. These are the most common uses of AI in business right now, and inherently the least risky; process improvement is the approach that most executives will consider first when deciding on their initial steps into the world of AI.

I've discussed case studies where processes can be streamlined through the application of AI Search (extracting meta-data from unstructured documents, for example) or by using big data to carry out predictive analytics (more accurate preventative maintenance schedules, for example). Other examples of process improvements are where AI has helped to filter CVs for job candidates, or helping make more efficient and more accurate credit decisions.

Whilst making existing processes faster, better and cheaper can provide a great deal of value to a business, there is arguably even more value available through **transforming the processes or the function**. By transformation I mean using AI to do things in a materially different way, or in a way that wasn't even possible before. Some of the examples of transformation that I have discussed earlier include: analysing the customer sentiment from hundreds of thousands of interactions (NLU); predicting when a customer is going to cancel their contract (Prediction); recommending relevant products and services to customers (Clustering and Prediction); predicting demand for your service (Clustering, Optimisation and Prediction); or modelling different risk scenarios (Optimisation).

One transformation that AI is particularly suited for is the enablement of customer or employee self-service. The benefits of a self-service option are that it can be made available twenty-four hours a day, seven days a week, and that it is generally cheaper to run. It also gives the customers or employees a sense of empowerment and control. AI can be used for the direct engagement aspects of the process, employing chatbots and/or speech recognition capabilities to communicate with the person, and also for any decisions that need to be made based on that communication (e.g. should this request for credit be approved?) by using prediction or reasoning tools. (Also useful when creating a self-service capability is RPA, which is able to handle all of the rules-based processing and connect all of the necessary systems and data sources together in a non-disruptive manner, as discussed in Chap. 4.)

The biggest impact that AI can make on a company is when a whole **new product, service or business** can be created using the technology at its core. Probably the most famous example of this is Uber which uses a number of different AI technologies to deliver its ride-hailing service. For example, it uses AI to provide some suggestions as to where you might want to go to, based on your trip history and current location (e.g. I usually want to go straight home after I am picked up from the pub). It will also help predict how long it is going to take until the driver arrives to pick you up. Uber have also analysed the time it takes to actually pick people up (i.e. from the car arriving at the site to when it leaves again) to enable the app to suggest the most efficient pickup spots. (After using third-party navigation apps, Uber eventually developed its own navigation capability, although it still relies on data from third parties.)

Other types of businesses have been created from a foundation of AI. (I'm not considering AI software vendors or AI consultancies here as they will inherently have AI at their core.) Powerful recommendation engines have been used by companies such as Netflix and Pandora to transform their businesses, and Nest, a 'smart thermostat' that uses predictive AI to manage the temperature of your house; without AI Nest would just be another thermostat. Other businesses that have initially used AI as a foundation for their core business are exploiting the technology to create new revenue. Pinterest, the website where people post interesting images from other sites, is a good example of this. They developed a very strong image recognition system so that users could find, or be recommended, similar images. They are now developing apps based on the same technology which can automatically detect multiple objects in images and then find similar images of the objects on the internet, including links to buy those objects (from which they are bound to take some commission).

So, understanding your AI ambitions is an important early step in developing an AI strategy and subsequent AI capability. Those ambitions will guide the first steps that you take and help steer you on the overall journey. But do bear in mind that not every part of your business will want or need to go on the same journey or at the same speed. Your front office might want to adopt AI wholesale whilst the back office is happy to take things slowly. And each department or function might be starting with different levels of automation maturity. If there are functions that have already adopted some form of automation or have made efforts to organise their data, these will provide stronger platforms to introduce AI than somewhere that still operates very manually or doesn't have a strong data governance.

The next section introduces the AI Maturity Matrix that can help you assess both the current automation maturity as well as your ambitions for each significant area of your business.

Assessing Your AI Maturity

A Maturity Matrix is a relatively simple concept that is a very useful on a number of fronts. Not only does it encourage discussion and reflection during the creation process but, once completed, it can also be used as a communication tool.

Maturity Matrices were originally developed by Carnegie Mellon University to assess the maturity of IT development functions. They generally describe five levels of maturity, from very immature (Level 1, or 'Initial') to world class (Level 5, or 'Optimising'). Each maturity level consists of related practices for a predefined set of process areas that improve an organisation's overall performance. Thus, an organisation achieves a new level of maturity when a system of practices has been established or transformed to provide capabilities the organisation did not have at the previous level. The method of transformation will differ at each level, and requires capabilities established at earlier levels. Consequently, each maturity level provides a foundation on which practices at subsequent maturity levels can be built.

Capability Maturity Model (CMM) levels in the IT development world can be formally assessed by approved consultancies, and many big IT companies wear their CMM Level 5 badges with pride. But not everyone should want or need to reach Level 5—in many cases Level 3 ('Defined'), where processes are documented, standardised and integrated into an organisation, is perfectly adequate for most businesses.

This idea of evaluating what is 'good enough' for your business, and each of the major functions in it, is at the heart of developing the AI Maturity Matrix. The methodology I describe in this section takes the concept of the Maturity Matrix and applies it specifically to automation. As I mentioned when discussing the Automation Strategy generally, it is best to try and think holistically about automation but with a strong focus and lens on AI, just so that related opportunities, or dependencies, are not missed along the way. So, in the matrix we are considering 'automation' that includes the full range of AI tools (chatbots, search, data analytics, optimisation engines, image classification, voice recognition, etc.) as well as RPA, robotics, IoT and Crowd Sourcing (which are described in detail in Chap. 4).

The AI Maturity Matrix has six levels rather than the usual five—this is to introduce a Level 0 where there is no automatson at all. Each of the six levels is described thus:

Level 0: Manual Processing There is very little evidence of any IT automation in the organisation. Only basic IT systems such as email and 'office' applications are deployed. There are large numbers of people processing transactional work, either in-house or through an outsource provider. Data is not considered an asset and there is no formal governance in place to manage it. There are no significant projects looking at, or delivering, automation.

Level 1: Traditional IT-Enabled Automation The organisation has implemented task-specific IT applications for particular processes (e.g. an invoice processing application to process invoices). There is no evidence of automation tools, and specifically AI nor RPA, having been deployed. Data is managed only to the extent that it needs to be to ensure the smooth running of the organisation. There are still large numbers of people carrying out transactional and cross-system processing work.

Level 2: Isolated, Basic Automation Attempts Some individual teams have used scripting or macros to automate some tasks within a process or application in isolated areas of the business. Minimal benefits are evidenced from these efforts. There are still large numbers of people carrying out transactional and cross-system processing work. Data is still managed on a nominal basis.

Level 3: Tactical Deployment of Individual Automation Tools Some functions have deployed individual automation tools such as AI and RPA to automate various processes. Some useful benefits have been identified. No dedicated automation resources are used or have been set up. Some data is managed so as to make it valuable for the automation efforts, but an organisation-wide data governance framework does not exist. In areas that have experienced automation, the staff are working in different ways. A case study or two from the organisation is known within the industry.

Level 4: Tactical Deployment of a Range of Automation Tools Functions or divisions within the organisation have deployed a number of different automation tools, including AI, across a range of processes. A strong business case exists that identifies sizeable benefits. Data is being managed proactively, with some areas having introduced data management and retention policies. Staff may have been redeployed as a result of the automation efforts, or they are working in materially different ways. Some dedicated resources have been organised into an automation operations team. The organisation has a reputation for implementing new technologies and being innovative.

Level 5: End-to-End Strategic Automation The organisation has implemented a strategic end-to-end process automation program with a range of automation tools including AI. Significant benefits in cost, risk mitigation and customer service have been delivered. Data is treated as a valuable asset and is managed through an organisation-wide data governance framework. Many of the organisation's staff are working in materially different ways than before. An Automation Centre of Excellence has been established. The organisation is renowned for its innovative and forward-thinking culture.

In the above descriptions, I have used the word 'organisation' to describe the scope of the work being evaluated—this could indeed be the whole organisation but could equally be applied to parts of it, such as the Finance Department, the Customer Services Department or a Shared Service Centre. When choosing the level of granularity to assess, it should be at the highest point (i.e. the biggest scope) that has a significant difference in automation approach and needs. Generally, this will be by function or division, but could also be by geography.

The evaluation of maturity levels is a subjective one, assessed through interviews and the review of evidence. The approach should be one that is satisfactory to all people who are involved and must be as consistent as possible across the different areas. Many people use a third-party consultancy to carry out the evaluation to ensure consistency but also provide independence from any internal politics there may be.

There can be some overlap between the different levels. For example, a department could have tactically deployed a range of automation tools (Level 4) but also created an Automation Centre of Excellence (a Level 5 criteria). It will be up to the person or team evaluating the levels to decide the most appropriate level to assign it to. The important thing is to be consistent.

As well as assessing the current automation maturity, the 'automation ambition' should also be agreed upon. As mentioned earlier, the target level doesn't necessarily have to be Level 5, and could be different for different areas that are being assessed. There are many reasons why not all areas should strive for Level 5, including the cost of implementation, there not being enough relevant data, it not being aligned with the strategic objectives and it simply not being appropriate (a human touch might be the best one in some cases).

A completed AI Maturity Matrix would therefore look something like the following (Fig. 6.2):

It shows both the 'as is' level of maturity (in grey) and the agreed 'automation ambition' for each area (in black). Although it is a relatively simple chart, it lays the foundations for the Automation Strategy and Roadmap. It also provides a useful communication tool to describe, at a high level, what the business is hoping to do with automation and AI.

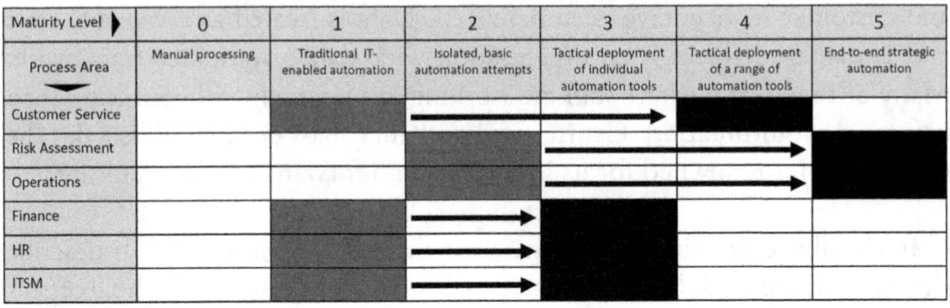

Maturity Level ▶	0	1	2	3	4	5
Process Area ▼	Manual processing	Traditional IT-enabled automation	Isolated, basic automation attempts	Tactical deployment of individual automation tools	Tactical deployment of a range of automation tools	End-to-end strategic automation
Customer Service		■	→————————————→			
Risk Assessment			■	→————————→		■
Operations			■	→————————→		■
Finance		■	→————→	■		
HR			→————→	■		
ITSM			→————→	■		

Fig. 6.2 AI maturity matrix

Levels 0, 1 and 2 do not involve the sort of automation that I am considering in this book. Level 3 is where AI (and RPA) are starting to be introduced, generally focusing on one or two individual processes. This equates to the 'ticking the AI box' ambition discussed in the previous section. Level 4 applies AI to a wide range of processes, applying it across a number of areas, although still tactically. This level equates to the 'process improvement' ambition. Level 5, where AI is applied strategically across the business, is equivalent to the 'transformation' ambition.

The ambition I described of creating a new service line or business by exploiting AI goes beyond the maturity matrix structure as it is not based on an existing function or department. That level of ambition means starting with a blank sheet of paper.

The AI Maturity Matrix then is a useful tool to start your AI journey. It provides an opportunity to openly discuss the role and opportunities for AI and automation across the business and provides a platform to develop a "heat map' of what those opportunities might be. The approach to creating an AI heat map is described in the next section.

Creating Your AI Heat Map

Following the development of the AI Maturity Matrix, the next step in creating an AI Strategy is to build a 'Heat Map' of where the opportunities for automation might be. Driven by the Business Strategy and the benefits that it is hoping to achieve, the Heat Map provides a top-down perspective on areas where AI is desirable, economically viable and/or technically feasible. It starts to identify the types of AI capabilities that could be applied in each area in order to realise the automation ambitions (and therefore contribute to delivering the business's Strategic Objectives).

The approach to creating an AI Heat Map is not too complicated as long as you understand your business well enough, and, of course, have a sound grasp of the AI Framework. It is intended as a starting point, a way to bring some focus and logic to prioritise your initial AI efforts.

Firstly, you must decide on the overall scope of the Heat Map. It usually makes sense to keep this consistent with the Maturity Matrix. So, if you originally evaluated the whole business split down into, say, five different business areas, then use the same structure for the heat map.

Each area, then, is assessed in turn, identifying the opportunities and the relevant AI capabilities required in each one. This is best done in two passes, the first one identifying all opportunities without any judgement being applied to them, rather like the brainstorming of ideas: all potential opportunities should be logged, dismissing none of them at this stage.

The opportunities are identified by considering a number of different criteria, each of which is described below. The opportunities are surfaced through interviews with relevant managers in the areas being assessed, and should be done by people who have a sound understanding of the AI capability framework as well as the AI technology market. Organisations without the appropriate resources in-house will tend to use third parties for this exercise.

- **Alignment with strategic objectives**—unless you are looking to implement AI for the sake of it, it is important to identify AI opportunities that contribute in some way to achieving the Strategic Objectives. For example, if your only strategic objective is to improve customer satisfaction, then the focus for this exercise should be on those processes that can impact customer service, rather than, say, cost reduction.
- **Addresses existing challenges**—there may be opportunities for AI to solve existing problems, such as inadequate management information, poor compliance or high customer churn. These opportunities may or may not be aligned with the strategic objectives but they should still be considered for inclusion as they could certainly add value.
- **Available data sources**—because AI, in most of its forms, requires large amounts of data to be most effective, a key consideration is to identify relevant data sources. Where there are very large data sets, then it is possible that there is an opportunity for AI to extract some value from them. Conversely where there is very little, or no data, it may preclude using AI. (Some AI technologies, such as chatbots and cognitive reasoning engines, both discussed in Chap. 3, do not require large data inputs, just captured knowledge, to work effectively, so don't simply equate low data volume to no AI opportunities.)

- **Available technology**—understanding the AI capabilities and the relevant tools that can deliver them is also important in order to identify AI opportunities. Whilst it is always best to 'pull' ideas in based on what the business requires, with AI at the forefront of technology it is also valid to 'push' ideas based on the technology that is available.

By this point you should have a range of different opportunities identified in each of the respective areas you are investigating. For example, in Customer Services, you could have identified the opportunity to provide an online self-service capability for the purchase of tickets for events. This could help deliver your strategic objectives to improve the customer experience and to increase ticket sales. In terms of the AI capabilities it could require Speech Recognition (if you want to offer telephone access), NLU (for the chatbots) and Optimisation (to guide the user through the purchasing process). It could also require some RPA capability to actually execute the purchase. (I would also include Crowd Sourcing, discussed in Chap. 4, as another capability that should be considered in order to support any of the AI opportunities.) With regard to data sources, there are pre-trained voice services available, and there is a good source of knowledge available from the human customer service agents that currently process these transactions (Fig. 6.3).

For each of these ideas, you should therefore understand what the opportunity is, how it links to strategic objectives (and/or addresses current issues) and which AI capabilities you will require to realise it. Building each of these on top of each other then starts to create a heat map of AI benefits and requirements.

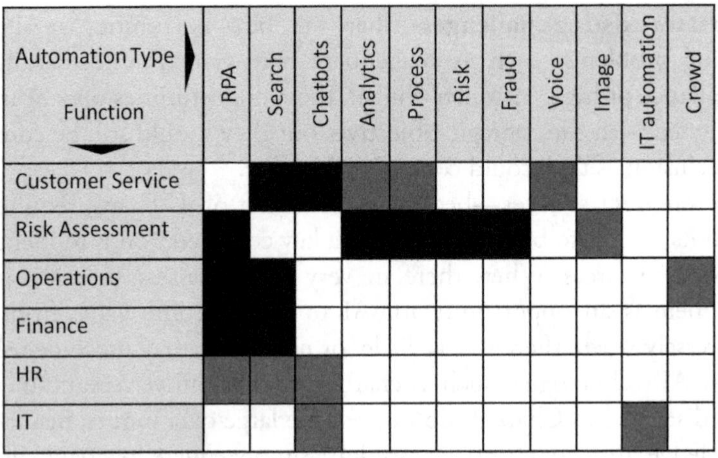

Automation Type ▶ / Function	RPA	Search	Chatbots	Analytics	Process	Risk	Fraud	Voice	Image	IT automation	Crowd
Customer Service		■	■	■	■			■			
Risk Assessment	■			■	■	■	■		■		
Operations	■	■									■
Finance	■										
HR		■	■								
IT			■							■	

Fig. 6.3 AI heat map first pass

(I tend to use colours to identify the 'hottest' areas, but you can use numbers or whatever approach you are used to.) Areas that are delivering the most benefit are clearly visible, as well as those AI capabilities which are going to be most important to you (Fig. 6.4).

The second pass across the various opportunities is to filter them down to those that are desirable, technically feasible and economically viable. If the idea can't satisfy all of these criteria then it is unlikely to be successful.

Desirability is a measure of how much the business wants and needs this new idea, so it relates to how aligned it is with the strategic objectives and how it addresses existing challenges. But it should also consider the opportunity from the customer's perspective (if it impacts them) and, culturally, how acceptable the idea will be to the department or function that will be implementing it. In some cases, you will have to consider the personality of the managers and staff of those areas as well and understand how supportive (or defensive) they may be. Clearly, if an idea does not pass the desirability test (to whatever threshold you think is appropriate for your organisation), then it probably shouldn't be progressed any further for now.

Technical feasibility will already have been assessed to a certain extent during the first pass when the data sources and the availability of technology were taken in to account. For the second pass this is done in more detail, considering aspects such as the quality of the data, the processing power or bandwidth that may be required, the maturity of the required technology and the technical skills that will be required internally and externally to implement it. Other aspects such as regulatory restrictions, especially about the use of data, may also have to be considered.

Testing the **economic viability** is the first look into what the business case might be like. A comprehensive business case will be developed in time, but at this stage the economic viability can be assessed at a relatively high level. It should start to consider the financial benefits to the business which could be achieved through cost reduction, risk mitigation, improved debt recovery, new revenue generation or reduced revenue leakage. Estimates can be made as to how the opportunity might improve Customer Satisfaction (CSAT) scores, if relevant. Costs should also be assessed where they can be at this stage. These could include license fees, IT infrastructure and professional services fees. It may be difficult to assess the costs at this early stage; therefore, it could be appropriate to score, out of 10, say, the magnitude of the costs, with anything scoring above a certain threshold rejected (this obviously needs to be balanced against the benefits that will be gained).

Each idea should be able to pass all three tests if it is to survive into the final Heat Map. You can, of course, make exceptions, but be clear on why you are

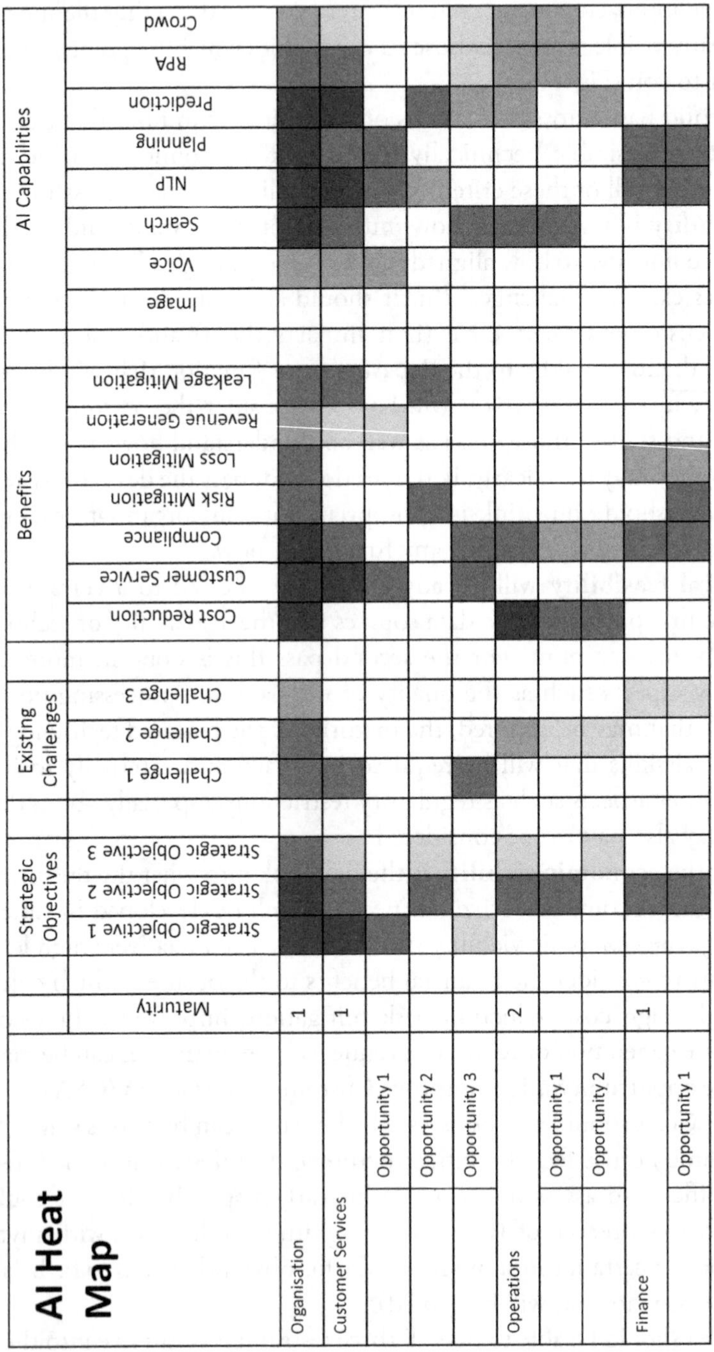

AI Heat Map

	Maturity	Strategic Objectives			Existing Challenges			Benefits							AI Capabilities							
		Strategic Objective 1	Strategic Objective 2	Strategic Objective 3	Challenge 1	Challenge 2	Challenge 3	Cost Reduction	Customer Service	Compliance	Risk Mitigation	Loss Mitigation	Revenue Generation	Leakage Mitigation	Image	Voice	Search	NLP	Planning	Prediction	RPA	Crowd
Organisation	1																						
Customer Services	1																						
Opportunity 1																							
Opportunity 2																							
Opportunity 3																							
Operations	2																						
Opportunity 1																							
Opportunity 2																							
Finance	1																						
Opportunity 1																							

Fig. 6.4 AI heat map

making those exceptions, and keep them in mind as you progress forward. Of the rejected ideas, do keep them safe. Things change, especially around the technological viability, so something that may not be suitable now may be just right in the future.

So now you have an AI Heat Map which shows the main areas you will be focusing on, the benefits they can bring and the capabilities required to achieve them. Each opportunity, as well as the high-level view presented in the Heat Map, should also have a corresponding passage of text (or a separate slide) which provides more of the necessary details about what the opportunity actually entails and some explanation of the numbers.

As a summary, and for presentation purposes, the AI Heat Map can be rolled up into an overview across the main areas and the organisation as a whole.

Now that we have a good idea of the opportunities that form the basis of the Automation Strategy, the next stage is to develop those Heat Map opportunities into a Business Case.

Developing the AI Business Case

Creating a Business Case for an AI project is, in many ways, the same as for any technology project—there are benefits to be achieved and costs to incur. But with AI the task is made more challenging because there tends to be more unknowns to deal with. These unknowns can make calculating an ROI (Return on Investment) an almost-futile exercise, and the organisation must rely on an element of intuition as well as the numbers. This is especially true when new approaches and new services are being developed through AI.

Luckily, the work done creating the AI Maturity Matrix and Heat Map provides a good methodology for evaluating which are the AI opportunities that will deliver the most value. Creating a meaningful business case for AI for each of these is certainly feasible with a little thought.

We already have, from the AI Heat Map, a list of opportunities and a high-level assessment of which are the most promising. In my example in the previous section the opportunities which are showing as the "hottest' (dark grey) are the ones to prioritise, especially if they are strong across a number of different aspects (strategic alignment, solving current challenges, benefits types). At this stage, you can easily apply some scoring mechanisms to make the prioritisation process easier. For example, you could replace the shades or colours with scores out of 3, and give weightings to each of the criteria, and then add all these together to give a total score for each opportunity.

If you are feeling confident, or have a third-party supporting you in this exercise, you could also introduce additional criteria under the broad heading of 'Ease of Implementation'. This would be influenced by the maturity of the function (from the Maturity Matrix) but also considers the technical feasibility and desirability that we looked at when filtering the opportunities during the second pass of creating the Heat Map. Scoring each opportunity in this way, again with appropriate weightings, will give additional depth to your decision-making process.

From this work, we now have a prioritised list of AI opportunities to consider: it's time to start getting serious with a few of these. Whether you take just the top scoring idea, or the top 3 or top 10 is entirely up to you, and will be based on your overall AI ambitions and the time and resources you have available. However many you take, it is likely that you will need to draft in some specialist resources at this stage, either from your own organisation or from third parties.

Different organisations have different ways of calculating business cases. Some will focus on ROI, some Net Present Value, some internal rate of return and others payback. I'm not going to explain each of these here as they are pretty standard approaches, and I am assuming that you are familiar with the one that is favoured by your own organisation. But all of these do require calculating the benefits and the costs over time, and so I will provide an aide memoire of what sorts of things you might need to include in each of these areas.

The benefits that AI can deliver are significant and diverse, and can be split into those that are 'hard', that is, can easily be equated to monetary value, and those that are 'soft', that is, those that are more intangible and difficult to quantify. It should also be noted that a single AI implementation may deliver a range of different benefits, not just one specific type. This is an important point—whilst you may implement an AI system to, say, reduce costs, there could well be associated benefits in compliance or risk mitigation. For each opportunity identified in the AI Heat Map, consider whether each of the benefit types described below could be applicable.

The Hard Benefits can be categorised as follows:

- **Cost reduction**: this is the simplest benefit to quantify as there is usually a base cost that can be compared to a future, lower cost. For AI, these are generally the situations where the new system will replace the humans doing a role or activities (remember the discussion in Chap. 1 around replacement versus augmentation?). The AI search capability will replace the activities of reading and extracting meta-data from documents—it will do it faster and more accurately than humans which means the benefits will be magnified. Chatbots (using NLU) can also replace some of the work done by human call centre agents, as can speech recognition when combined with a cognitive reasoning engine. NLG (a sub-set of NLP) can replace business analysts in creating financial reports. More efficient route planning for workers and/or vehicles reduces time and cost.

- **Cost avoidance**: for businesses that are growing, cost avoidance is the more acceptable face of cost reduction. Rather than recruiting additional people to meet demand, AI (as well as other automation technologies such as RPA) can be implemented as an alternative. The AI solutions will be similar to those for cost reduction.
- **Customer satisfaction**: this is usually measured through a survey-based satisfaction index such as CSAT or Net Promoter Score (the difference between positive and negative customer feedback). Some businesses have connected this to a monetary value, and others relate it directly to a manager's own evaluation and bonus. For many it is a key performance indicator. AI has the opportunity of improving customer satisfaction through being more responsive to queries, being more accurate in the responses, providing richer information in the responses, reducing the friction in customer engagements and in making relevant recommendations. (Of course, AI can also be used to automatically measure customer satisfaction itself through sentiment analysis.)
- **Compliance**: some people might challenge the fact that a self-learning system can improve compliance which is based on hard and fast rules, but actually AI is very good at identifying non-compliance. Using NLU and Search it can match policies and procedures against regulations and rules and highlight the variances. Compliance benefits can be measured through the potential cost of not being compliant, such as through fines, or loss of business in specific markets. (RPA, by the way, is a strong driver of compliance as each automated process will be completed in exactly the same way every time, with each step taken logged).
- **Risk mitigation**: AI can monitor and identify areas of risk in areas where it would be impossible, or very expensive, for humans to do it. The classic example for AI here is fraud detection in high-volume transactions, such as credit card payments. AI can, in some circumstances, make better risk-based decisions than humans, such as for credit approvals. AI can also contribute to the Know Your Customer (KYC) process in being able to validate documents and data sources and help with the credit checking process (RPA also helps here to access the necessary systems and run the process).

As with compliance, risk mitigation can be measured by the costs that have been avoided, such as the loss through fraud or poor credit decisions.

- **Loss mitigation**: loss mitigation can be achieved through improved debt recovery. Most of the heavy lifting here can be done through RPA by managing the timing and activities in the process, but AI can play a role, for example, in generating appropriate letters or engaging with the debtors. Loss mitigation can be measured through the increase in cash being recovered through the use of automation.

- **Revenue leakage mitigation**: this is usually achieved through reducing the situations where revenue generating opportunities are lost, such as when customers no longer use your services. Reducing customer churn through the early identification (i.e. clustering) of the tell-tale behaviours can be measured by calculating the revenue value of a customer. Other AI capabilities, such as NLU and prediction, can also contribute by keeping the customer more engaged with your business.

- **Revenue generation**: this is the AI benefit that can probably deliver the most in terms of monetary value. AI can help enable self-service capabilities (through Speech Recognition, NLU, Optimisation and RPA, for example), identify cross-selling and up-selling opportunities (through Clustering) and create new revenue generating opportunities either from your existing business or through completely new services or products. Revenue generation is easy to measure after the fact, although isolating the specific contribution of the AI may be challenging and assumptions will need to be made. Predicting revenue generation will need to use modelling techniques, most of which should already be available in your business.

One thing to bear in mind with measuring automation benefits is that, in some cases, staff productivity could actually be seen to decrease after implementation. This is not a bad thing, but just a reflection that the staff are now handling the more complex cases whilst the technology is handling the simpler ones. Overall, the productivity will have improved, but if you just look at the human staff they may well be taking longer to process the (more complicated) cases.

The softer benefits are inherently more difficult to put a monetary value to, but can still be evaluated, even if the accuracy is less certain:

- **Culture change**: this is probably the most difficult thing to achieve and the most difficult to measure, but can deliver significant value to an organisation. Of course, it depends on the type of culture you have already and what you would like to change it to, but AI can help embed a culture of innovation as well as deliver customer centricity into the business. Actual benefits will usually be associated with a wider change program and it will be challenging to isolate these that are specifically due to AI, but, because of the potential value, these should not be ignored.

- **Competitive advantage**: this can deliver significant benefits if AI provides a first-mover advantage, a new service line or access to new markets or customers. The benefits will be closely associated with (and usually accounted for in) some of the 'hard' ones such as cost reduction, customer satisfaction and revenue generation, but competitive advantage should always be sought as a way to deliver a step-change in value.

- **Halo effect**: AI, at least in the current decade, is a highly marketable aspect to have within a business. The objective of being able to 'tick the AI box' can provide the opportunity to claim you are an 'AI-enabled' business which could attract new customers to your product or service. Any amount of AI implementation, if marketed correctly, can also have a positive impact on shareholder value if it results in an increase in the share price of the company.

- **Enabling other benefits**: as well as providing direct benefits, AI can also enable other, indirect benefits to be delivered. For example, implementing AI can free up staff that can be deployed on higher-value activities, or data that is generated from an automated process can be used to provide additional insights that can be exploited in another part of the business. Implementing AI may, through eliminating much of the mundane work in a department, also reduce employee churn.

- **Enabling digital transformation**: transforming businesses to be more 'digital' is a common strategic objective for many businesses right now. AI is clearly part of that transformation, and many of the benefits of a digital business can be directly delivered, or indirectly enabled, through the implementation of AI.

An AI business case will include at least one of the above hard and soft benefits I have listed and often quite a few of them. The individual classes of benefits should be used as a starting point for your AI business case. For each AI opportunity identified in the heat map, consider each type of benefit carefully so that all can be identified and none are forgotten or left out.

Once all the relevant calculations have been done, the benefits for each opportunity can be summed across the departments and the organisation to give a full picture of the value that AI can bring. Also, consider whether there are any synergies between the different opportunities—for example, does having a number of AI initiatives going on enable a favourable cultural change in that department?

The heat map, bolstered with the business case data, now provides a comprehensive prioritised list of AI opportunities within your business. But, before you take the first step into prototyping and implementing the most favourable candidates, you will need to understand what your ultimate roadmap looks like and how you will handle all the change management aspects that go along with it. These two pieces of the AI strategy jigsaw are covered in the next two sections.

Understanding Change Management

Even the simplest IT projects require change management. With automation and AI, the challenges are magnified—not only are you going to fundamentally change the way people work, but you may also have to let people go. And even if this is not your intention, the reputation of automation projects means that staff will likely think it is the case.

You may, as part of your AI strategy, be creating a new product or service that will be competing with your legacy products or services. That 'healthy competition' may be good for business but is not necessarily good for staff morale. And building a new capability will mean transforming your organisation—new jobs will be created and old jobs will disappear. Some are going to be able to cope with the change, and some are not.

Right at the beginning of the book I talked about the differences between AI replacing humans and AI augmenting them. Clearly an augmentation approach will have less change management issues than a replacement one. But anything that can demonstrate how AI will enrich the work that people are doing (even if they have to do it differently) will help. Having greater insights into their work, or customers, or suppliers, through AI analysing and interpreting data, will help people become more successful in their jobs.

If automation is replacing tasks, it is often for the more transactional, tedious ones. There is a useful phrase coined by Professors Willcocks and Lacity, which is that "automation takes the robot out of the person" (Willcocks and Lacity 2016). So, if your AI projects in any way relieve people of the mundane, repetitive tasks that they have been doing for years, then this should be celebrated.

Despite the amount of change that AI inherently throws up, many of the common change management practices are relevant to AI projects: communicate early and often; involve and empower employees as much as possible;

look for short-term wins; consolidate gains to generate further momentum; incentivise employees where appropriate; pace the change appropriately and have honest reviews before the next one.

A challenge that automation projects in particular face is a post-prototype slump. The initial build can become a victim of its own success—the relief of actually creating something that works, even if it is not a complete solution, can be so overwhelming that everyone then takes a step back, resulting in a loss of focus and momentum.

To overcome this, it is first necessary to be aware of it, and then to prepare for it. Schedule in review sessions for as soon as the prototype is due to be completed. Have a Steering Group Meeting soon after, or get the CEO to make a celebratory announcement about how this is 'the end of the beginning'. If possible, have the next phase of the project, or another prototype, start before the first project finishes, so that focus can quickly shift to that. The worst thing you can do at this point is to pause for a few weeks or months.

Generally, a 'big and bold' approach to an AI program will deliver the most success, as it avoids any potential slump periods, and helps maintain that all-important momentum. However, all organisations are different and an 'under the radar 'approach may be the most suitable for you—just make sure that your business case and roadmap (covered in the next section) reflect that.

As well as managing all the change within your organisation, you may have to manage change for your customers as well. If your AI project is one that impacts the way customers engage and interact with you, then you will need to fully prepare them for that change. This will involve plenty of communication, supported by marketing efforts and relevant collateral such as emails, website articles and frequently asked questions (FAQs).

If your AI project changes the way that you use your customers' data, then you will need to make it clear to them what the impact of this is. Any sensitivities around data privacy, in particular, will need be made explicit. You would be well advised to seek legal advice for any changes to your customers' data usage.

This isn't a book about how to manage change (there are plenty of those about already), and you will likely already have your own preferred methodologies to tackle this. This is only to say that you should not neglect change management in any way—implementing AI brings about its own unique challenges, especially when some of the changes in an organisation can be seismic in nature. These challenges need to be understood and carefully managed, and built into your AI roadmap.

Developing Your AI Roadmap

The AI Roadmap provides a medium- to long-term plan to realise your AI Strategy. It can be a relatively simple affair that draws heavily from the AI Maturity Assessment and the AI Heat Map.

The Maturity Assessment will provide you with your starting point. Those areas that are immature when it comes to automation will require more effort and time to complete. Your Change Management Assessment will also direct you to those areas that might need additional support along the way.

At the core of the Roadmap will be the opportunities identified through the AI Heat Map and subsequent Business Case. But rather than having a series of individual opportunities being worked through, it is good practice to group them into common themes. These themes can represent a number of different projects, or work streams, ideally centred around common AI tools. For example, a process excellence stream might focus on improving the accuracy of processes and reducing the AHT, whilst an analytics and reporting stream would focus on delivering compliance and improved reporting. Each stream, using a range of different technologies, relates specifically back to a number of benefits that are outcomes of the business strategy.

The AI Roadmap should be kept at a relatively high level, identifying the specific project work streams and associated activities (change management, program management, governance, etc.). For each work stream there should then be a much more detailed project plan which identifies dependencies and responsibilities (Fig. 6.1).

It is likely that you will want to build one or two prototypes of your most favourable candidate opportunities, and these should be built into the roadmap, ideally with their own detailed project plans. More details on the approach to prototyping are provided in the next chapter. The prototype builds will test and validate assumptions but also provide important momentum for your AI program.

The key considerations when building a roadmap include: your ultimate ambition with regard to automation (is this a game changer for your business or just something you want to try out in certain areas); whether you want to evaluate the whole business first and then implement AI or try and get some quick momentum by implementing in some specific areas first; and how much change your organisation can bear at any one time (see previous section).

It is important to remember that your AI Roadmap will be an important communication tool. It should set out clearly what you are going to do and when, so that you can use it to explain how it will deliver your AI strategy, and ultimately contribute to your overall Business Strategy.

Creating Your AI Strategy

Your AI Strategy is now effectively complete. You will have looked at your Business Strategy and worked out what aspects of that could be satisfied by the application of AI technologies. You will have determined how ambitious you want to be with your AI program—whether it is just to be able to say you have some AI, or to improve some processes, or to transform parts of your business. You may have even decided to create a new business by exploiting AI.

You will then have looked at your organisation and assessed its level of AI maturity. Some functions or departments could still be working manually, whilst others could have started some automation projects and others might even have some AI projects in place already. For each of these areas you will have worked out their AI ambition, and thus determined the size of the gap that you have to work with.

The AI Heat Map is the tool that helps identify how you are going to bridge that gap. Based on your knowledge of the AI Capabilities Framework you will have identified a number of different AI opportunities in each of the areas, assessing and prioritising their potential benefits, challenges and their alignment with the Business Strategy.

From the information in the AI Heat Map you will have developed a high-level business case that sets out both the hard and soft benefits that you expect to achieve.

And finally, you will have considered the change management activities that are necessary to ensure the success of the AI program, and built them into an AI roadmap that sets out your medium- and long-term plans of how you will realise the AI strategy.

It is now time to start building something.

The Cognitive Reasoning Vendor's View

This is an extract from an interview with Matthew Buskell, Head of Customer and Partner Engagement at Rainbird, a leading vendor of Cognitive Reasoning software.

AB: Why do you think AI is being so talked about right now?

MB: I think it's a combination of forces coming together and driving the interest. Specifically Outsourcing, Cloud Computing, Big Data and, of course, Hollywood.

The major driving force for outsourcing, realistically, is economic. However, regulation has increased and margins have continued to reduce. So, when outsourcing does not give a deep enough return we need to look elsewhere. In this case AI is attractive because it allows us to amplify the productivity of the staff we already have in a way that is not possible by outsourcing alone.

Cloud computing is a big part of the reason AI has become economically viable. That and the algorithms that have been optimised and refined over the past 30 years. To illustrate this let me tell you about Ian. In 1994 Ian was doing undergraduate studies in software engineering and AI with me at Birmingham. He decided to code a PhD thesis on linguistics. When he ran his program the processing power required was so great that it ground the entire Sun Hyperspark network of servers and workstations to a halt for the entire campus. So he was banned from using the computers outside of the hours of 10 pm–6 am. As a result, we did not see Ian for nearly a year. Today we call his field of research Natural Language Understanding and you can buy it for a fraction of a cent per hit from several cloud providers.

Big Data has made AI interesting. For some types of AI like Machine Learning to work well they love large amounts of data. Whilst it's true most businesses were collecting data 20 years ago, until recently they had not put the big data platforms in place that would allow AI systems to access them. Now that they do, AI is well placed to take advantage of it.

Hollywood might seem a weird one to add, but that fact is as humans we have been fascinated by the idea computers could think since Ada Lovelace described the first computer in the eighteenth century. TV programs like Channel 4's *Humans in the UK* and movies like *Ex-Machina* fuel that excitement. It also sells news articles and ad-words on blogs. There is a famous, but anonymous quote that sums up Hollywood's influence on AI: "Artificial Intelligence is the study of how to make real computers act like the ones in the movies."

This also poses a big challenge for the AI industry today. The fact is we are way off simulating a human mind, so the expectation of AI is so high it is bound to lead to some disappointments along the way.

AB: What value do your customers get out of your software?

MB: The Rainbird software is about mimicking human decisioning, judgement and recommendations. It's different to other AI in that its 'people down' rather than 'data up'. Meaning we take a subject matter expert and sit them down with Rainbird so they can teach it.

When you stop for a moment and think about what you can do if you mimic human expertise and decisioning the list is endless. We have been involved with projects from Abbey Road studios trying to mimic the expertise of their (rather ageing) sound engineers, through to large banks and tax consultancies trying to reduce headcount by encoding the best employees' expertise into Rainbird.

On the whole, however, we have noticed two areas where we see significant value:

1. If you can take expertise and create a solution that allows customers to self-serve you can drive some staggering efficiencies. At the moment, this is being manifested as a 'chatbot'. However, long term, we believe companies will have a 'cognitive platform' and the UI may or may not be a chatbot.

2. If you can take knowledge that was in a process and move it further upstream the impact to the downstream processes is dramatic. In one instance, we moved the liability resolution in a motor claim so that it could be handled by the call centre agent. This removed the need for over 15 downstream processes.

AB: What do customers or potential customers need to focus on if they are going to get the maximum value from AI?

MB: I have noticed a lot of companies rushing into AI proof-of-concepts without spending enough time working though the customer journey or the business case.

There are some customers that are approaching this correctly. For example, I was with a client the other day in Scotland that was taking a much more Design Thinking approach to the development of a PoC. When we started the process we all had the idea we could get up and running quickly with a simple bot that could answer FAQs. When we actually looked at the questions and followed the conversation with a real agent it quickly became clear that the general advice would only last a few seconds before it became specific. At that point, you needed a human in the loop, so the result of putting AI in place is that you would have paid twice—once for the virtual agent and secondly for the human. This is a Bad Idea. We were actually able to find the value by going deeper with the virtual agent and make it capable of completing a full interaction with a smaller number of questions.

AB: Do you think the current hype around AI is sustainable?

MB: Personally, I don't think the hype is helpful, so I welcome it slowing down or becoming more grounded. However, I think unfortunately it will continue, mainly because we need AI to work. Without it we have to face some pretty stark economic realities.

AB: How do you see the market developing over the next 5 years or so?

MB: The market is very fragmented at the moment and the use cases for the technology are very broad. So, I think it will do what software innovation has always done: all these companies will consolidate so they can offer an end-to-end solution and the use cases that work well will survive and the rest will die out.

　　The final thought I would leave you with is from John McCarthy, who is considered the father of AI: "As soon as it works it's no longer called AI anymore."

Reference

Willcocks LP, Lacity MC (2016) Service automation: robots and the future of work. Steve Brookes Publishing Warwickshire, UK.

7

AI Prototyping

Introduction

Creating your first AI build, however small, is a key milestone for any AI program. After all of the hard work developing the AI Strategy, this is the point at which things really start happening, and people in your organisation can actually see tangible outputs. Getting this stage right, therefore, is vitally important.

There are a number of different approaches to this, which I cover in some detail in the 'Creating Your First Build' section of this chapter, and they each have different names and acronyms. For simplicity, I have used the word 'prototype' as the general term for all of these.

Prototyping can be done once the AI Strategy is complete, or can be done to validate some of the assumptions during the development of the strategy. This may be influenced by how important those assumptions are (if they are fundamental to the success of the program, then test them early) and how much stakeholder buy-in and momentum you need (a successful prototype is a great way to engage with stakeholders).

Some of the decisions you make at the prototyping stage may have an impact on the rest of the program. The most important of these is your technology strategy—do you build your AI solutions yourself, buy them from a vendor or build them on top of an established AI platform, or use a mixture of all of these? It is this question which I will address first.

© The Author(s) 2018
A. Burgess, *The Executive Guide to Artificial Intelligence*,
https://doi.org/10.1007/978-3-319-63820-1_7

Build Versus Buy Versus Platform

Some of the business case costs that were discussed in the previous chapter will have required some assumptions on how you intend to create your AI solutions. Generally, there are three main approaches to take: off-the-shelf AI software, an AI platform or a bespoke AI build. Your final solution may involve any of these approaches, but the core functionality is likely to initially focus on one of these:

- **Off-the-shelf AI Software** is generally the simplest approach, since most of the hard work in designing the system has already been done by the vendor. As I've described in the earlier chapters, there is a plethora of AI vendors, each of which can provide a specific capability that can be used as a stand-alone application or as part of a wider solution.

    As well as all the design, testing, debugging and so on having already been done, implementing off-the-shelf software means that the product will be supported by the vendor, and they are likely to have either their own implementation resources, or partners trained up to support implementation. You will still need to exert effort to identify, clean and make available the required data, as well as providing subject matter expertise regarding your own processes (so 'off-the-shelf' is a bit of a misnomer), but generally this is the 'path of least resistance' for implementing an AI capability.

    The biggest disadvantage of using a software 'package' is that its capabilities may not align well enough with your objectives and required functionality. You may want to implement, for example, a system to analyse the sentiment of your customers' feedback. One vendor may provide a system that analyses the data exactly as you would have expected but doesn't provide any useful reporting tools. Another may be less effective at the analysis but has a best-in-class reporting suite.
- There are also a number of risks in relying on a software vendor to provide your AI capability. Because of the excessive hype around AI right now (remember Chap. 1?) there are quite a few vendors that are little more than an idea and a website. The challenge is that many of the AI software vendors are start-ups or young companies, so it can be a difficult task to identify those with viable and stable products. Your due diligence should be extremely thorough so that it filters out those who do not yet have enough experience or capability. Using AI experts at this selection stage can save considerable pain later on in the project. (You may, though, wish to deliberately go with a software vendor that does not yet have a fully commercialised system if their technology could potentially provide you with a distinct

competitive advantage, or where you want to help them shape their offering, so that it delivers some specific advantage to you, whether that be functionality, exclusivity or price.)

The commercial models for AI software can be quite varied. The most common, and simplest, is an annual subscription that includes the necessary licenses as well as the support and maintenance of the software. It will usually include hosting of the software by the vendor (or an agent of the vendor) if it is a software-as-a-service model. These AI software packages are fully formed, with a user interface for training and using the product. For other types of AI software where the software is accessed through an API (e.g. to determine the sentiment of a particular passage of text) a 'pay-as-you-go' model is used—the user pays per API call made. There are usually bands of pricing depending on the volume of calls made.

- **AI Platforms** are the middle ground between an off-the-shelf-package and a bespoke build. They are provided by the large tech companies such as IBM, Google, Microsoft and Amazon, some large outsourcing providers, such as Infosys and Wipro, as well as specific platform vendors such as H20, Dataiku and RapidMiner.

In the section on 'Cloud Computing' in Chap. 4 I described Amazon's 'stack' of AI services, which is typical of a platform offering. It provides ready-made, ready-trained algorithms for those with simple and well-defined requirements (such as generic text-to-speech translation); untrained but ready-made algorithms for those that need to add their own specific practices, nuances and data into the model; and a set of AI tools for researchers and data scientists to build their algorithms from the ground up. (The platform provider can also usually provide the infrastructure facilities as well.) Of course, an organisation may want to use a range of these services for the different capabilities that their solution requires.

The platform approach may make sense if your organisation already has a relationship with one of the tech giants, especially if there is already a strong connection with cloud services. You may find, though, that you have to make some compromises on certain aspects where a platform provider's specific AI capability may not be as strong as a stand-alone one. Of course, you can always mix and match between the two types, depending on how rigid your partnership model is.

Outsourcing providers also provide AI platforms for their clients—for example Infosys have Nia and Wipro have Holmes. They usually combine a data platform (e.g. for data analytics), a knowledge platform (to represent and process knowledge) and an automation platform which brings these together with RPA. If your outsourcing provider does have a robust AI platform then it is certainly worth looking at it—you will probably have to use their resources for the implementation, but it does provide a relatively simple route assuming that their capabilities match your requirements.

As well as the platforms provided by the tech giants and outsourcing providers, there are also development platforms available from AI-only vendors, such as H20, Dataiku and RapidMiner (these are sometimes referred to as Data Science Platforms). These provide a solid foundation for developing AI applications with your own resources or using third parties. They consist of a number of different tools and application building blocks that can be integrated with each other whilst providing a consistent look and feel. This approach will give you the greatest flexibility and control out of all the platform models—the tools are generally more advanced than those provided by the bigger platforms. You will need to take a more hands-on approach to the design and implementation though—the main users of these systems will be the data scientists.

The pricing models vary between the different providers. For commercial developments, some offer a revenue share model where they will charge an ongoing percentage (usually around 30%) of any subsequent revenue generated from the application built on their platform. The most common approach is to charge per API call—this means that every time the application that you have developed uses a specific piece of functionality of the platform (e.g. converting a piece of text to speech) there is a small charge. Commonly there is a threshold below which the API calls are free or at a fixed price.

In the platforms space, there is some overlap between each of the categories I have defined—IBM, for example, could easily fit into all three—but this does provide a useful framework for selecting which one (or ones) may be most appropriate for your AI strategy.

- **Building bespoke AI applications** will provide you with the greatest level of flexibility and control; however, it is unlikely that you will want to take this approach across all the opportunities that you have identified. Bespoke AI development, just as for any bespoke IT development, can be great for providing you with exactly what you need but can create related issues around change management and support. Therefore, for the majority of enterprises, bespoke AI development should be used only when absolutely necessary, that is, for complex, very large data problems, or when creating a completely new product or service that requires technological competitive advantage. (It may make sense to use bespoke development for creating the initial builds—see next section for details—but subsequently moving onto a vendor or platform approach.)

A bespoke build will also require effort to be put into designing and building the user interface for the AI system. This can be good in that you can develop something that suits your users perfectly, but can be a challenge to get it just right. Many excellent and clever AI systems have failed because the user interface simply wasn't good enough.

For a bespoke development, you will need your own highly capable data scientists and developers or (more likely) to bring in specialist AI consultancies that already have these resources. The commercial model is usually based around Time & Materials, although Fixed Price and Risk/Reward can also be used when the requirements are well defined.

You may want to use a combination of all the above approaches. If your AI strategy is ambitious and will impact many areas of the business, then you will probably want to consider the AI platforms as your prime approach, making use of off-the-shelf packages and bespoke build where appropriate. If you only want to target one or two AI opportunities, then it may make more sense to consider just the off-the-shelf solutions. A strategic program to create a brand-new product or service may require bespoke development to ensure that it delivers enough competitive advantage.

So, now that you have an idea of the development approach, or approaches, that you are going to use, it is time to actually build something.

Creating Your First Build

At this point in your AI journey you will have a list of prioritised opportunities, each with an estimate of the value that it could deliver, as well as an idea of how they could be implemented (build, buy or platform-based). For the next stage, we need to test some of our assumptions and build some momentum for the program, which means we have to start building something.

The scale of this initial build stage will depend very much on your intent. There are five generally accepted approaches to this, each of which focuses on satisfying slightly different objectives. You won't need to understand the detailed intricacies of each of these approaches (your internal development team or external partner will have the necessary knowledge) but it is important to know which is the most appropriate approach for your project so that you don't waste valuable resources and time.

There are some overlaps between the different approaches, and you could legitimately choose specific aspects from each of them. You may also want to use a number of these approaches in turn, as they each focus on different objectives to be achieved.

- **Proof of Concept (PoC)**: a PoC is a software build that will be discarded once it has been able to demonstrate that the concept that is being tested has been proven. It can be of any scale, but generally focuses on the barest functionality whilst ignoring any complexities such as security or process exceptions (this is called 'keeping to the happy path'). PoCs will not go into live production, and will use non-live 'dummy data'. PoCs are usually carried out to test the core assumption (e.g. we can automate this process using this data set) but are also useful to manage stakeholder expectations—seeing a system in the flesh early on often provides an extra boost of momentum for the project as a whole.

- **Prototyping**: this is a broad term that covers the building of specific capability in order to test the viability of that capability. Prototypes may focus on the user interface (usually called horizontal prototypes) or specific functionality or system requirements (vertical prototypes). One project may have a number of different prototypes, each testing for a different thing. It is rare for a prototype to resemble the final product. As with PoCs, prototypes are generally discarded once they have done their job, but some can be incorporated into the final product (this is called 'evolutionary prototyping')—it is important to know at the start whether the prototype will be retained as it will need to be built more robustly than a throwaway one.

- **Minimum Viable Product (MVP)**: the proper definition of an MVP is one where the build, which contains the minimum level of functionality for it to be viable, is released to its customers or users. The feedback from these early adopters is then used to shape and enhance the subsequent builds. The term MVP has, though, become a catch-all description for any sort of early build and therefore tends to lose focus as to what it might be looking to achieve. For a build that is not intended to be released, then a RAT (see below) is generally the better approach. For a build that will go live, then a Pilot (again, below) is usually preferred. MVPs can have their place in the development cycle, but only once their objectives have been well defined.

- **Riskiest Assumption Test (RAT)**: RATs are a relatively new approach that focus on testing only the riskiest assumption that requires validation. This is in contrast to the MVP which is less specific about what it is trying to achieve. Also unlike MVPs, RATs are not intended to go into production. A RAT will seek to build the smallest experiment possible to test the riskiest assumption that you have. As with prototypes (RATs are really a specific type of prototype) a number of different tests are carried out to validate different assumptions—once the riskiest assumption has been tested and validated, then the next RAT will focus on the next riskiest assumption and so on. Understanding what the riskiest assumption actually is requires some detailed thought—you will need to identify all of the assumptions that you have made so far that need to be true for the AI opportunity to exist. Ask yourself if these assumptions are based on other assumptions, then try to identify the root assumptions and then find the riskiest one of those. Then it is a case of working out how you can test that assumption with the minimum amount of code. RATs are not the easiest approach to take but can deliver the best long-term value for your project.

- **Pilot**: a Pilot, like an MVP, will be used in a production environment. Unlike an MVP, it should be fully formed and complete. It is considered as a test because it focuses on automating a process or activity that is relatively small and low risk to the organisation. Pilots take longer to build and are more expensive than any of the options above, but have the advantage of not 'wasting' time and effort (although focused prototyping is rarely a waste of time). The Pilot needs to achieve all of things that are expected of a PoC and MVP as well as being fully functional; therefore, the candidate process needs to be chosen well—if it is too complex (e.g. it involves lots of systems or has many exceptions) then it will take too long (and cost too much money) to build, and crucial momentum will be lost. A successful Pilot will provide a huge confidence boost to the project as well as deliver tangible value.

As with the wider automation strategy, when selecting the most appropriate approach for that initial build always bear in mind what your end goal is—ask yourself whether you need to build any sort of prototype at all (the answer is probably yes) but also which approach is going to provide you with the highest chances of success. Speed of implementation and cost will also be big factors in the choice of your approach. As a default answer, the RAT is likely to be the best approach for the majority of cases, but do consider all possibilities.

Understanding Data Training

One aspect where AI projects are generally trickier than 'normal' IT projects is with the dependency on data, and this challenge is particularly acute during the prototyping stage.

Although it is possible to cut down on the functionality and usability of the system being built in order to test assumptions, it is rarely possible to do the same with the training data. Of course, not all AI systems rely on large data sets—cognitive reasoning engines and expert systems require knowledge rather than data, but for machine learning applications in particular, data availability is paramount.

There will be a minimum amount of data that will produce a viable outcome, and this can sometimes mean that the prototype will have to rely on nearly all the data available, rather than a small sub-set. If your complete data set contained, say, 10 million records, then a sample size of 10% may still be workable, but if your complete data set was, say, 100,000 records then (depending on what you intend to do with it, of course) a sub-set may not give you any meaningful results.

The more complex the model you want to create (i.e. one with many different parameters), the more training data you will need. If the data is insufficient, your AI model could suffer from what data scientists call over-fitting—this is where the algorithm models the fit of the data too well and doesn't generalise enough (it 'memorises' the training data rather than 'learning' to generalise from the trend). To understand whether your model is exhibiting over-fitting and correct for it there are various statistical methods, such as regularisation and cross-validation, that can be exploited, but usually at the expense of efficacy.

A good portion of your training data must also be used for testing. This validation set should, like the training set, be 'known'; that is, you should know what output you are expecting from it ('this is a picture of a cat') but you don't reveal this to the algorithm. Because AI outputs are probabilistic the answers won't always be correct, so you need to understand and agree what level of confidence you are comfortable with. There is no hard and fast rule about how much of your training data should be retained to do this testing but 30% is a good starting point. Factors such as the size of the complete set, the richness of the data and risks of getting the wrong answer will impact this figure.

You may not be able to get around the fact that, in order to train and test your prototype system effectively, you need to use nearly all of the data you have available. If data availability is going to be a challenge then one or more RATs, as described in the previous section, may be the best way forward at this point.

Once you have a good understanding of how much data you will need for the prototypes, you will need to acquire the data, (probably) clean it and (maybe) tag it. I can't stress enough how important the quality of the data is to the value that the AI system will deliver.

The most likely scenario is that the data will be sourced from within your own organisation—it may be customer records, transaction logs, image libraries and so on. You may also want to acquire data from external sources such as the open-source libraries I discussed in the section on big data in Chap. 2. This data will come ready-tagged, but there may also be publicly available data sets (such as traffic movements) that will require tagging before they become useful for training your algorithm.

The cleanliness, or accuracy, of the data will clearly have an impact on the efficacy of your AI system. The old adage of 'garbage in, garbage out' is particularly relevant here. Common sense and system testing will determine whether your data is of sufficient quality to provide useful results.

For training data that needs tagging, such as images or video, the most efficient way is through crowd sourcing (see Chap. 4 for how crowd sourcing works). Crowd sourcing firms, such as CrowdFlower, have many human agents at hand to complete micro-tasks—each are sent items of the data and simply tag them with the required information (this is a picture of a cat, this is a picture of a dog, etc.). The fully tagged data set is then returned to you so that you can train the model.

One risk with crowd sourcing is that any biases in the humans doing the tagging will eventually be reflected in the AI model. As a rather silly example, if the human agents doing the tagging were all very short, and they were asked to tag pictures of people as to whether they were tall or not, then it may be that a higher proportion than expected would be tagged as tall. That 'smallness bias' would be carried through into the model and skew the results of the new, unseen data. Most good crowd sourcing firms though will ensure that bias doesn't enter the data in the first place, but can also correct for any bias that may exist.

Another, rather innovative, approach to data acquisition and training is to use virtual worlds or computer games instead of 'real life'. I mentioned previously that Microsoft have created a specific version of Minecraft which can be used to train AI agents, and some software companies are using games like Grand Theft Auto to help train their image recognition software that is used in driverless cars. So, rather than having to drive around many different environments for hours and hours filming and then tagging objects, all of it was done in a fraction of the time in the virtual world of the computer game.

So, with your AI Strategy complete, and your first build underway (whichever way you have decided to approach it), it is time to think about the longer term, and how you can start to industrialise your new-found AI capability. But not before getting a good understanding of all the risks that AI can raise, and how you might mitigate these, and that is the subject of the next chapter.

The AI Consultant's View

This is an extract from an interview with Gerard Frith, co-founder of Matter and Sensai.ai, AI Strategy and Prototyping consultancies.

AB: What got you into the world of AI in the first place?

GF: I have been fascinated with the mind since I was small child. I read Jung and Freud in my early teens, Godel Escher Bach when I was 18. I initially studied Psychology at university, but in my first year it was my Cognitive Science module that grabbed my attention most. I became obsessed! At the end of my first year, I decided to start over, and signed up for an AI degree instead. That was 1991.

Once my degree was done, I found out that the rest of the world was rather less interested in AI than I was. During the AI winter of the 1990s, I mostly had to pretend my degree was just an exotic computer science degree.

In 2013, the rest of the world was finally catching on, and I established Matter AI, one of the first AI consultancies anywhere. I sold Matter in late 2016.

My major focus at the moment is on the use of AI to create better and deeper customer relationships.

AB: Why do you think AI is being so talked about right now?

GF: There are two key reasons: firstly, AI tech has recently become practically useful. That's mostly due to the vast amounts of data that the digital age has exponentially spewed, and to the massive advances in available processing power. The theoretical advances in AI that underpin much of what is being used in business today were developed years, and sometimes decades, ago.

The second reason is that AI has always been exciting, scary, troubling and challenging in a way that no other technology is or could be. If you take away the breathless discussions of how AI will take all our jobs, then you are left with a pretty appropriate level of coverage of a disruptive technology.

The creation of genuine AI could easily claim to be the most significant scientific achievement in history. It has the potential too to be one of the most significant social and spiritual achievements.

AB: What value do your customers get out of your services?

GF: My customers say they get three main things from working with us:

- Expert guidance. I've been a developer, a CEO, and a management consultant, so I know the technology, I know the market and I understand what a company needs to be successful.
- I'm a disruptor. I like to challenge the current equilibrium and bring new thinking to situations that allows organisations to reinvent their value propositions and the way they deliver them to market.

- Because of my breadth of experience, I can provide the glue that brings together tech and business strategy, and then how to execute those strategies in effective ways.

AB: What do customers or potential customers need to focus on if they are going to get the maximum value from AI?

GF: The same things they should always focus on when assessing new technology—how can I create greater value for my customers using this technology? What competitive advantage could this give me? The critical thing to do is to move fast. It's getting easier for disruptors to change industry dynamics quickly, and with less capital. But do treat AI as an early stage technology too. Invest seed capital in experiments, but get those experiments live quickly and then iterate.

AB: Do you think the current hype is sustainable?

GF: It's inherent in human systems that hype is never sustainable. Our biology seeks novelty and in doing so makes cycles of hype and deflation inevitable. That said, I do believe that AI is worthy of tags like the '4th Industrial Revolution' in ways that, say, the 2nd and 3rd ones weren't. In this sense it's just a continuation of the impact of technology development in general, but its impact will be quicker, but only in the same way that the internet's impact was quicker than that of TV, and the impact of roads occurred faster than the impact of the printing press.

Beyond its vast (but normal) economic impact though, I think it will actually change everything for humans. It is not like any other technology in terms of how it affects how humans will think about themselves. Currently even professors of neuroscience who argue that the mind is nothing greater than a complicated computer hold on privately to the idea that they have free will and a 'soul'. AI will challenge those ideas in quite new and scary ways.

AB: How do you see the market developing over the next 5 years or so?

GF: The next five years will be very exploratory. AI is a general-purpose technology and will thrust out in multiple directions at once. We are seeing new products appear weekly covering a plethora of use cases, with even the biggest tech players only covering relatively small areas. Companies will need to look at gluing together a range of best of breed providers, often with a strong vertical industry focus.

We'll undoubtedly see a lot of acquisitions, but only slow consolidation as tech companies focus on land grab. We'll also begin to see some industries such as Law begin to be taken over by software. In general, industrial boundaries will become even more porous as data drives the discovery of new horizontal processes that can cross the traditional verticals.

8

What Could Possibly Go Wrong?

Introduction

It is important to provide a very balanced view on the benefits and risks of AI. Throughout the book, I've talked about some of the unique challenges that need to be faced when implementing AI, but it is worth drawing these out into a separate chapter, as I have done here, to make sure they get the attention they need.

In the chapter I've tried to cover the 'localised' challenges—those that any company will have to consider when implementing AI—but also the more general ones that will affect the way that we work and how we will live our lives.

The Challenge of Poor Data

Data quality is traditionally measured by the '4 Cs': Clean, Correct, Consistent and Complete. But in the world of AI and Big Data we need different rules to guide us. Data quality can be thought of as a combination of data veracity (its accuracy) and data fidelity (its quality in the context of how it is being used).

Data veracity is less important in the world of Big Data because any small errors will be drowned out by the sheer volume of correct data. The model that is created from the algorithm will look at overall trends and fitness, so a few outliers will not significantly impact the outcome. (If there are fundamental errors in the data, such as the decimal point being one place out on every data point, then, of course, it will have a material impact, but in this section I am only considering small issues in data quality.)

© The Author(s) 2018

A. Burgess, *The Executive Guide to Artificial Intelligence*,

https://doi.org/10.1007/978-3-319-63820-1_8

With data veracity, whether the data is being used for AI or not, there is always a balance to be made between accuracy and cost. The trick is to understand what level of accuracy is 'good enough'. With AI applications, that threshold is generally lower than for more traditional computing applications because of the generalised modelling effect.

Data fidelity, on the other hand, can have a serious impact on the results. This is where the data is inappropriate for the task for which the AI has been set. In Chap. 5 I wrote about one example of AI being used to predict recidivism of people who have been charged with criminal offences by the Durham Police Force. I noted that the data used for the predictions was sourced only from the region itself, and therefore excluded any relevant activity the subject might have taken part in outside of Durham County. So, a career criminal from Leeds (just down the A1 main road from Durham) who had just committed his first offence in Durham may have been allowed to walk out of the police station because he was considered a low risk. In this case, the data that the AI system is using is correct but not wholly appropriate.

Another, slightly different, reason that the source data would not be appropriate is if it contained unintended bias. This is a specific problem with AI systems and is covered in detail in a later section in this chapter.

Related to data quality is the question of whether more data always means better results. Generally, with machine learning applications this is true; however, there are some subtleties to consider. There are two, almost conflicting, perspectives.

The first, called 'high variance', is where the model is too complex for the amount of data, for example where there are many features compared to the size of the data. This can lead to what is called over-fitting and would cause spurious results to be returned. In this case, the data scientist could reduce the number of features, but the best solution, and the one that seems most like common sense, is to give the model more data.

But for the opposite case, where the model is too simple to provide meaningful results, called 'high bias', adding more data to the training set will not deliver any improvement to the outcome. In this case adding more features, or making the available data a higher quality, both in veracity and in fidelity (such as data cleansing and outlier removal) will be the most effective approaches to improving the outcome.

Generally, it is the role of data scientists to ensure that the data is big enough, relevant and appropriate for the problem at hand. Although my police example above is relatively simple, some aspects of data science can seem more like an art form than a science. In most cases there is no clear 'right' and 'wrong' answer—the data scientist must use creativity and inference to identify the best sources and mix of data. This particular skill set is one of the reasons that the best data scientists can demand such high fees at the moment.

Once poor data quality has been identified as an issue then there are a number of approaches to try and fix (or enhance) it. Some of these could be the traditional methods of data cleansing; crunching through the data to identify anomalies to constraints and then correcting them using a workflow system.

Crowd Sourcing, which I described in Chap. 4, can also be used as a cost-efficient way to cleanse or enhance data sets, particularly with regard to ensuring labels are correct in training data sets. Tasks, usually split down into small units, are assigned to a distributed team of humans, who will process each one in turn.

RPA can also be a useful tool to help cleanse data where large enterprise workflow systems are not available. The robots will need to be configured to identify all the constraints they need to look for. Once they have found the data errors they can then validate those with other systems to determine what the correct answer should be, and then make that correction. RPA can be a good solution for 'one off' cleansing exercises because it is relatively easy to configure. It can, of course, be used in combination with crowd sourcing.

And, with only a slight element of irony, AI systems themselves can be used to cleanse data for AI systems. Some vendors now have solutions which can carry out Field Standardisation (e.g. making two different data sets consistent), Ontology Mapping (e.g. extracting product characteristics), De-duplication (e.g. removing entries with the same content) and Content Consistency (e.g. identify and matching abbreviation forms). Generally, they work by using machine learning to analyse the structure of the data model to then determine the kind of errors such a model is likely to generate.

Another area where AI could be used to improve the quality of data, and even generate new data, is in the use of Generative Adversarial Networks (GANs). This is where one AI system judges the outputs from another system. For example, the originating AI could generate a picture of what it thinks of as a cat. The second system then tries to work out what the picture is of. The assessment, i.e. how close is this to looking like a picture of a cat, is then used to guide the first system to try again and create a better picture. GANs are relatively new, and this approach is still at the experimental stage, so I will save a detailed description of how this works to the final chapter.

As should be clear from much of this book, AI (in most cases) feeds off data—without the data then there is little value that AI can add. But, as we've seen in this section, more data does not always mean better results (especially in high bias model scenarios). And it's not just about volume of data but the quality. The veracity and fidelity of the data can have a huge impact on the performance and effectiveness of the outcomes. Any data-dependent AI project must therefore involve careful planning, ensuring the data is the appropriate size and complexity for the problem that is being addressed.

Understanding the Lack of Transparency

The reason Machine Learning is called Machine Learning is, rather obviously, that it is the machine, or computer, that does the learning. The computer does all the hard work in 'programming' the model that is going to make the predictions. All it needs is the data to be trained on: feed it the data and it will come up with the model. The human (the data scientist) has to choose the right algorithms in the first place and make sure the data is appropriate and clean (see previous section), but the model that is created is essentially the work of the machine.

I labour this point somewhat because it is an important, inherent aspect of machine learning that is responsible for one of its biggest drawbacks—the lack of transparency. I can ask a trained AI system to, for example, approve a credit lending decision, or to recommend whether a candidate gets short-listed for a job, but I can't easily ask it how it came to that decision. The model has looked at the many different features of that customer or that candidate and, based on the many different features of the training set, has determined the answer to be, in these examples, yes or no. But which features were influential in that decision, and which ones were irrelevant? We will never really know. It's a bit like trying to recreate an egg from an omelette.

Which, of course, can be a problem if you need to explain to that candidate why they weren't short-listed, or why you didn't give that customer the loan. And if you are in a regulated industry, then you will be expected to be able to provide those answers as a matter of course.

There are three general approaches to tackling AI opaqueness. The first one is to not use machine learning in the first place. In Chap. 3, in the section concerning the optimisation capability, I described Expert Systems and Cognitive Reasoning Systems, which are particular flavours of AI that work on the principle of a human subject matter expert creating the original model rather than it being created using data. (For this reason, some AI experts claim that expert systems should no longer be described as AI.) The 'knowledge map' that is created from these approaches can be interrogated, by a person or via a chatbot, to access the information. In some of the more advanced systems, the queries can be automatically routed through the map using the specified weightings that have been applied to each node and connector.

This all means that the decision that comes out at the end can be traced back through the map, making it fully auditable. So, using one of my examples above, the failed credit application could be traced to, say, 75% of the decision due to salary, 40% due to postcode, 23% due to age and so on. And, once it has been designed, assuming the processes don't change, it will remain consistent in its decisions.

The challenge in using expert systems is that they can get very complex very quickly. The more features that the system has, the more complicated the model becomes to define. If the processes do change, the model then has to be changed as well to reflect them, and that can be a laborious task.

Another approach for simple problems is to use self-learning decision trees, usually called Classification and Regression Trees. From a set of input data, these CART algorithms create a decision tree that can then be interrogated to understand which features have been the most influential. CART alogorithms generally provide the best balance between effectiveness and transparency.

For complex systems with lots of data and many features, an 'ensemble' approach that calls on many different algorithms to find the most popular answer (Random Forest is the most common ensemble approach), will prove the better choice, but it will have that challenge around transparency. The most common approach to tackling this is to try and reverse-engineer the decision by changing just one variable at a time. Returning to my customer who has had his loan rejected, we could repeat his case but changing each of the features that he has fed into the system (his salary, his postcode, his age, etc.). This trial-and-error approach will then be able to indicate (but not necessarily isolate) which of the features had the most influence on the decision.

The problem with this approach is that, with many features, this can be a long process. Carrying out the analysis on a number of test cases and publishing these first is a good way to avoid analysing every rejected case. The 'model cases' could demonstrate, for example, how, with everything else being equal, age impacts the decision-making process. But you might have to do that for lots of different model cases since the influence of age might be different in different postcodes or for different salary levels. And, if the system self-learns as it goes along, then the test model will need to be refreshed at regular intervals.

The level of testing required will be very dependent on the type of problem that is being solved. The more sensitive the problem (medical imaging, for example), the more robust the testing will need to be.

A disgruntled customer who has just been rejected for that loan may, then, be satisfied with this demonstration of the 'logic' used to make that decision. But would that satisfy the industry regulators? Again, each use case will have different requirements for testing, but it could be the case that being able to demonstrate a fully tested system, and the influences that different features have, will be enough for some regulators. In the legal sector where e-Discovery software is used to automatically process thousands or even millions of documents to identify those that are relevant or privileged in a litigation case, the law firm using the software must be able to demonstrate a robust use of the solution, with due process applied at every stage. Once the court is satisfied that this is the case, then the results from the analysis can be used as valid evidence.

Artificial Intelligence is a new and fast developing technology—regulators are generally slow and considered in their approach, and will need time to catch up. Putting the risk of fraud to one side, once AI is more generally accepted in society it may be the case that customers, and therefore regulators, take a more relaxed approach to the transparency question. Until then, AI-adopting companies need to ensure that they have the answer when the question is raised.

The Challenge of Unintended Bias

A common fallacy about AI systems is that they are inherently *unbiased*: being machines they will surely lack the emotional influences that inflict humans in our decision making that lead to bias, whether intended or not. But that is just a fallacy. The core of the problem is that the data used to train the AI may have those biases already baked in. If you use biased data to train an AI, then that AI will reflect back those same biases. And, because the model can be opaque to interrogation (as we saw in the previous section), it can be difficult to spot them.

Here's a simplified example to demonstrate the problem. In the world of recruitment, AI can be used to short-list candidates based on their CV (curriculum vitae, or résumé). The AI system would be trained by feeding in many CVs, each one labelled as either 'successful' or 'unsuccessful' based on whether they got the job they were applying for at the time. The AI can then build up a generalised picture of what a successful CV looks like, which it can then use to short-list any new candidate CV that is submitted—a close enough match between the candidate CV and the 'successful' model would mean the CV gets through to the next round.

But how did those candidates in the training set get their jobs? By being assessed by humans, with all our conscious and subconscious biases. If the human recruiters had a propensity to reject older candidates (even if they didn't realise they were doing it), then that bias would flow through into the AI model.

So, how to ensure there is no bias in the training set? The first answer is to try and use as wide a sample as possible. In my example above, you would try to take training CVs from many different recruiters, ideally those with as neutral a mix of gender, race, age and so on as possible (just as polling firms try to use a representative sample of the population).

This may not always be possible, especially if the data is from your own company and is all you have. But even publicly available data sets can be prone to bias. Some of the data sets used for facial recognition training do not have enough representative samples from people of colour. This means that any AI trained using that data could struggle to identify non-white faces. As these biases in public data sets are unearthed, they are being slowly fixed (there is even an Algorithmic Justice League which highlights bad practices like these), but AI developers should be particularly cognisant of the issues with any public data that they use.

In any of the above cases where there could be bias in the data, it will be necessary to try and test for it. Just as in the previous section, where we looked at testing for different influences in decision making, we can similarly test for the influence on any particular feature for bias, as long as we know what the 'right', that is, unbiased, answer should be.

In the recruitment example, we would expect an unbiased response to show that age does not influence the short-listing decision. If we are able to isolate the age feature in our data set then we can test, by varying only the age in our model case, whether it is influencing the outcome at all. As with the transparency test, this would need to be done across a range of samples (e.g. just men, just women) to ensure that age was not influential for any particular group.

Assuming some level of bias has been detected, the next question is what to do about it. Finding the type of bias (e.g. age, gender) will immediately help identify any groups of data in the training set that may be causing it (there may have been a sub-set of the data that was from a particularly young set of recruiters, for example). This input data can then be excluded or altered to eliminate the source of that bias.

The model can also be tweaked by changing the relative weightings between different terms. If the training CVs in my example showed a strong correlation between Manager (role) and Male (gender), the mathematical relationship between gender-neutral words like 'Manager' and gendered words such as 'Male' can be adjusted. This process, called de-biasing, is not insubstantial, and usually involves humans to identify the appropriate and inappropriate words, but it can be done.

One important thing to bear in mind is that not all bias is necessarily bad. There may be instances where the inherent bias is important and where the consequences justify the means, such as in identifying fraudsters. There is clearly a fine balance between representing the truth, biases and all, and altering data to represent the accepted social position. Also, on a much simpler level, the AI will need to understand specific definitions in areas that could be biased, such as in the difference between a king and a queen, if it is to work properly.

Eliminating unintended bias from AI is, largely, still a work in progress and must be considered carefully where there are specific consequences. Using publicly available data sets does not, currently, ensure neutrality. Using data from your own organisation will make de-biasing even more important, and therefore must be built into the AI development roadmap.

Understanding AI's Naivety

It was the French philosopher Voltaire who famously said that you should judge a person by the questions they ask rather than the answers they give. These may be very wise words when it refers to humans, but for AI the situation is even simpler: the machine doesn't need to know what the question is in the first place in order to give a respectable answer.

This is the whole idea behind Clustering, which I described in Chap. 3. You can present a large volume of data to a suitable algorithm and it will find clusters of similar data points. These clusters may depend on many different features, not just a couple of things like salary and propensity to buy quinoa, but, in some cases, many hundreds. The AI is providing mathematical muscle beyond the capability of a human brain to find these clusters.

But these clusters are not (or, more accurately, don't have to be) based on any pre-determined ideas or questions. The algorithm will just treat the information as lots of numbers to crunch, without a care whether it represents data about cars, houses, animal or people. But, whilst this naivety of the data is one of AI's strengths, it can also be considered a flaw.

For big data clustering solutions, the algorithm may find patterns in data that correlate but are not causal. In the section on Clustering in Chap. 3 I gave the rather whimsical example of an AI system finding a correlation between eye colour and propensity to buy yoghurt. It takes a human to work out that this is very unlikely to be a meaningful correlation, but the machine would be naive to that level of insight.

The AI may also find patterns that do not align with social norms or expectations—these usually centre around issues such as race and gender. I've already written in the previous section on the challenges of unintended bias, but in this case an awkward correlation of purely factual data may naively be exposed by the algorithm. The challenge for those responsible for that algorithm is whether this is a coincidence or there is actually a causality that has to be faced up to. How that is handled will have to be judged on a case-by-case basis, and with plenty of sensitivity.

There is also the infamous example of the Microsoft tweetbot (automated Twitter account) that turned into a porn-loving racist. It was originally intended that Tay, as they called the bot, would act through tweets as a 'care-free teenager' learning how to behave through interactions with other Twitter users. But it quickly turned nasty as the human users fed it racist and porno-graphic lines which it then learned from, and duly repeated back to other users. Tay, as a naive AI, simply assumed that this was 'normal' behaviour. It only took a few hours of interaction before Microsoft were forced to take the embarrassing tweetbot offline.

One useful way of thinking about the naivety of AI is to consider how dogs learn. Like all other dogs, my own, Benji, loves going for a walk. I know this because he gets excited at the first signs that a walk might be imminent. These include things like me locking the back door and putting my shoes on. Now, Benji has no idea what the concepts of 'locking the back door' or 'putting my shoes on' are, but he does know that when these two events happen in close succession then there is a high probability of me taking him for a walk. In other words, he is completely naive to what the preceding events mean—they are just data points to him—but he can correlate them into a probable outcome.

(This dog/AI analogy is quite useful and can be extended further: my dog is quite lazy, so if he sees me lock the back door but then put my *running* shoes on, he goes and hides to make sure I don't take him with me. In this scenario, he is using increased granularity to calculate the outcome this time—it's not just 'shoes' but 'type of shoes'. Of course, he doesn't know that my running shoes are specially designed for running, just that they are different enough from my walking shoes. It may be the different colour/shade, a different smell, the different place where they are kept and so on. This demonstrates the opaqueness issue I discuss in a previous section: I have no real idea (unless I do some pretty thorough controlled testing) what aspect of the shoes switches the outcome from 'Excellent, I'm going for a walk' to 'Hide, he's going for a run', but it clearly does have a binary impact. I should also point out that the dog/AI analogy also has its limitations: Benji has lots of other basic cognitive skills, such as knowing when it is time for his dinner without being able to tell the time, but because AIs are currently very specialised in their capabilities, an AI that predicted walks would not be able to predict dinner time.)

So, the naivety of AI systems can be a real headache for its users. Suffice it to say that the outcomes from the clustering must be used carefully and wisely if they are to yield their full value. Data scientists and AI developers must be aware of the consequences of their creations, and must apply heaps of common sense to the outputs to make sure that they make sense in the context for which they were intended.

Becoming Over-Dependent on AI

As you will have seen from the many examples provided throughout this book, AI can achieve some remarkable things, particularly when it is carrying out tasks that would be impossible for humans to do, such as detecting clusters of like-minded customers (or fraudsters) in databases containing many millions of data points.

The problem may come when companies become overly dependent on these systems to run their business. If the only way to identify your best customers or people trying to defraud you is through very complex AI algorithms, then there will always be a risk that these systems actually stop working effectively or, probably worse, stop working effectively without you realising it.

This all comes down to the complexity of the problems that are being solved and therefore the minimal understanding of how they are really working. I have already covered the lack of transparency of how an AI may come to a decision, but for very complex algorithms (and one AI solution usually contains a number of different types of algorithms all strung together) there will only be a few people—the AI developers and data scientists—that understand how these have been designed and built in the first place. Being dependent on a few (very in-demand) people is clearly a risk for any business. It's rather like the world of traditional IT where a core legacy system has been hand-coded in an obscure language by one developer who could leave the company, or get run over by the proverbial bus, at any time. The risk can be compounded further if the AI system is being built by a third party who will not necessarily have that long-term commitment to their client that an employee might.

There are mitigating actions though. Most of them are similar to the approach one would take with the bespoke legacy system coder—make sure that they have documented everything that they do, and are incentivised to stick around to help support the system. If they won't be around for long, make sure there is a robust succession plan in place.

For a third party, there can be more pressure applied, through contractual obligations, to get the solution fully documented, but the most important thing will be to select the right provider in the first place, which I discuss in more detail in Chap. 9. In that chapter I also write about setting up a Centre of Excellence (CoE) that will be the guardian of the necessary skills and documentation to maintain and improve the AI solutions in the future.

Another factor to bear in mind when trying to solve complex problems is that it is difficult to know whether the system is providing a correct, or reasonable, answer. For some AI solutions, such as measuring sentiment analysis, we can test the output of the machine with what a human would reasonably say was the correct answer; for example, the machine said the sentence 'I was very dissatisfied with the service I received' would be predominantly 'negative' and a human would agree with this. If there was a high volume of sentences to assess (as would usually be the case) then a sample can be tested. But for complex problems, such as trading financial instruments or designing drugs, it is nearly impossible to tell if the machine is making a correct decision. We can look at the final result (the end-of-day trading position, or the efficacy of the drug) but we will not be able to determine if this was the best outcome of all—perhaps more profit could have been made or a better drug could have been designed.

There is also a more philosophical risk from the over-dependency on AI: as AI becomes more and more commonplace in our lives, we will eventually lose the ability to do even the simplest cognitive tasks because we are no longer practising those skills. Our ability to remember people's names and 'phone numbers, and our ability to read maps are already being eroded by our dependency on smartphones and satnavs.

As AI capabilities become greater, they are bound to impact more of our cognitive skills. Some people may argue that this is not necessarily a bad thing, but we have developed all those skills over many millennia for particular reasons and, absent of the technology to help us, we quickly become exposed and vulnerable.

But, coming back to a more practical basis, AI skills will, over time, become more commoditised (just as HTML development skills have in the last 20 years) and the problem of dependency on highly skilled AI developers and data scientists will gradually recede. But for now, the risk of over-dependency should be built into the AI strategy as soon as it is clear that the solutions will go beyond an off-the-shelf or simple platform approach.

Choosing the Wrong Technology

The field of AI is fast moving—what was not possible in one year can be solved in the next. All the drivers for AI that I described in Chap. 2 are all continuing to get better and more influential—big data is getting bigger, storage is getting cheaper, processors are getting faster and connectivity between devices is now almost infinite. So, how do you choose which approach and technology to take today, if tomorrow something better could come along? (Users of Apple products, by the way, face this dilemma every time they think they want to upgrade anything.)

One of AI's constraints, that each application can only do one thing well, is actually an advantage here. If an AI solution has been built up from a number of different capabilities (as per the AI Framework) then each of those individual capabilities could be swapped out with a newer, better approach if one comes along. Replacing one chatbot solution with another will not necessarily impact any Optimisation capability you have already built. You will still have to train the new piece of AI but you won't necessarily have to throw away the whole solution.

Say, for instance, that you are using one of the AI platforms exclusively to provide all your required AI capabilities. If one of the other platform providers brings out a better AI capability for, say, text-to-speech, then it is not too big a challenge to switch out the current API connection for the new one. (Some providers will, of course, try to lock you into their platform through contractual means, or even just canny sales techniques, so you will need to watch this as you select your approach and provider.)

Bigger investments will, of course, require greater levels of due diligence. Putting in a multi-million-pound vendor-based AI solution will certainly commit you to a roadmap that may be difficult to step off. But that is true for any large IT investment.

It does become more challenging if there is a fundamentally new approach available. Established users of Expert Systems will have looked at Machine Learning when it first came along with some envy. But if we assume, as is likely, that machine learning (and all its associated approaches like DNNs) will be the fundamental core technology of AI for a long while, then it should be a relatively safe bet to base a strategy around. (The only technology that may have a material impact on machine learning is Quantum Computing, but this is still very much still in the labs and will take decades to deliver practical day-to-day uses.)

Probably the bigger issue with committing to one particular AI technology is the skills that you will need to retain to develop and support it. Generally, individual developers will be specialists in a particular tool or platform; therefore, changing the tools may mean changing the developers. If these resources are being bought in through a consultancy or implementation partner, then you will just need to make sure that that firm has capabilities across all the relevant tools and platforms that you may require (I cover this in more detail in the Vendor Selection section of Chap. 9). If you have started to build up a Centre of Excellence based around specific technologies though, you may have to think hard about whether the change is worth it.

Preparing for Malicious Acts

With great power (as Spiderman's Uncle Ben famously said) comes great responsibility. All the capabilities that I have discussed throughout this book have shown the huge benefits that AI can bring. But that power could also be used for nefarious means as well.

The ability for AI Clustering to, for example, identify customers that may buy a certain product, could also be used to identify people who are ideal fraud targets, especially if it is triangulated with other data sources. Criminal AI systems would usually have access only to publicly available data (which is always more detailed than you think) but, as has been seen with the numerous large-scale hacking attacks, other, more private data may also be available to them.

In early 2017, some banks introduced voice passwords for their online services—all the customer had to do was train the system on their voice and they could then login by simply saying 'My voice is my password'. Within months it could be proved that this approach can be fooled using facsimiles of the user's voice—the BBC showed in one example how a presenter's non-identical twin could log in to his account—and the worry now is that AI-powered voice cloning technology will be able to do the same job. The AI will need voice samples to be trained on, but, again, these are surprisingly common across social media, and if you are a famous person, then widely available. There is already a commercial mobile app, CandyVoice, which claims to be able to clone voices.

It won't just be banking systems that could be fooled by voice cloning technologies; you and I might think we are listening to our mother on the telephone saying she has forgotten the password on her bank account, or our boss asking us to send specific emails to suppliers or to make certain payments.

Similarly, Image Recognition can be used to subvert Captchas—these are the small images that appear when you try and make a payment on certain websites. You can proceed only if you can type the numbers that are in the photos, or identify only those pictures with, say, cars in. These are meant to avoid software robots making unauthorised purchases, but are now being bypassed by armies of people, or clever AI systems, entering the correct answers.

The difference between the type of online criminal behaviour we have been used to in the past and what we see now is the scalability of the AI. Voice and image recognition can be done at high volume and at low cost—if just a very small fraction get through then the criminals will be successful. It's the same numbers game that spammers exploit.

AI can also be used to socially engineer people's behaviours. During recent political campaigns (Brexit in the United Kingdom is one that comes to mind) software 'bots' were used to influence people's opinions through targeting their social media accounts with pertinent messages. Social media companies already use AI to change the order of the articles in a person's social media feed so that the most 'relevant' are closest to the top. External companies can use a person's online behaviour to predict which way they might vote and then try to reinforce or change that using targeted posts and tweets. Extending this to include more nefarious acts than voting is within the realms of possibility, especially if the accounts that are being posted from are pretending to be someone else.

Artificial Intelligence systems can also aid malicious behaviour by being trained to do the wrong thing. 'Dirty data' could secretly be fed into a training set to change the learnt behaviours of the AI. The large search engines are constantly having to defend themselves against malicious companies who create fake sites to increase their ranking in the search results.

In the earlier section on the issue of naivety in AI systems, I wrote about Microsoft's tweetbot, Tay. Tay didn't understand the nastiness of the 'data' that it was being fed, but Microsoft clearly didn't expect the deliberate sabotage of its creation by mischievous users. Although Tay is a slightly humorous example (unless you work for Microsoft, of course), it does show how AI can very quickly be influenced by dirty data. Extrapolate this to a situation where the AI is controlling millions of interactions with customers, or even financial transactions, and you can start to see the potential risks from this.

The fixes for the types of behaviours I have described above are diverse, and usually very technical (apart from insuring against loss, which should be done as a matter of course). Ensuring that your voice recognition application cannot be a victim of voice cloning will be an inherent part of its development. With Captchas, they need to be updated on a regular basis to ensure that the latest technology is being deployed. To stop dirty data getting in requires the usual defensive approaches to stop anyone from infiltrating your systems. The level of defences you put up will depend on the sensitivity of the data and the risk of anything going wrong.

Social engineering is more difficult to ameliorate. It is necessary to swim against an incoming tide of more and more people who want to use, and generally trust, social media channels. The mitigating actions are awareness and education. Just as young people (mostly) know that someone at the other end of a messaging conversation may not be who they claim to be, people also need to be aware that the chatbot at the other end of their chat conversation may not be who, or what, it claims to be.

As with all online activities, being one step ahead of the criminals is the best that can be hoped for. AI has the ability to do very good and useful things but, as we have seen in this section, also has the potential to do bad things at scale. Simply being aware of what can go wrong is the first step in combating it.

Conclusion

For readers hoping for a book all about the benefits and value of AI, this chapter may have come as a bit of a shock. But, as I said at the start, it would have been remiss of me not to include details of how AI can also present risks to businesses, by being either misappropriated or sabotaged.

I deliberately included some of the worst cases of those risks, as awareness of what can possibly go wrong will be the most important part of being able to mitigate them. Some of these defensive steps are part of the normal IT development cycle, some are unique to AI development and some are more sociological. For big, complex AI projects, you will be dependent on strategists, developers, security experts, data scientists and sociologists to ensure that your application is not open to abuse or unnecessary risk.

In Chap. 10, the final one, I will cover some of the more philosophical debate around the impact of AI, including how it affects jobs, and the big question of what happens when (if ever) AI gets smarter than us.

But, before that, we need to look at how your AI efforts and projects can be industrialised and embedded into your business so that they provide sustainable benefits for the long term.

The AI Ethicist's View

This is an extract from an interview with Daniel Hulme, CEO and Founder of Satalia, a London-based AI development firm with a strong ethical mission.

AB: Daniel, first tell me about Satalia, and how you approach AI.

DH: In this capitalistic world right now, businesses are naturally looking at AI to either make more money or to reduce costs. Satalia helps companies do this by providing full-stack AI consultancy and solutions, but our higher purpose is to 'enable everyone to do the work they love'. We think we can achieve that by building a completely new 'operating system for society' that harmonises technology with philosophy and psychology. We are taking the core learnings from the solutions we have developed for our clients and then making them available as blueprints and tools for people to use to make their businesses and lives better. Satalia is also a global voice for purposeful AI start-ups, of which there are not enough of in this world right now.

　　　Within Satalia itself, our staff are free to do what they want—they set their own salary, working hours and vacation days, and they have no KPIs. We combine AI and organisational psychology to enable our employees to work on the projects they want to, unburdening them from bureaucracy, management and administration. This gives them the freedom to rapidly innovate in ways that they would rarely find anywhere else.

AB: Can you talk to me more about how you see the ethics of artificial intelligence playing out?

DH: There are three perspectives of AI that need to be considered differently if we are talking about ethics.

　　　The first is where decisions are made through a trained but static model. The ethical question is who is responsible if that model doesn't behave according to our norms—if it is racist or sexist for example. This is the Unintended Bias problem. This is linked to the big challenge of building explainable algorithms—how do we have transparency from what is essentially a black box model? Now, if governments legislate for these things, will that stifle innovation? Or to stay competitive should companies be allowed to 'move fast and break things' as the current trend goes? AI developers have a real responsibility in the middle of all of this, and breaking things without careful consideration of the impact is not the best way forward.

Secondly, there are the more general AI solutions where multiple types of AI are combined into an ever-adapting system that changes its model of the world when in a production environment. The most famous example of this is the Trolley Problem that designers of driverless cars are facing. Should the crashing car hit the child or the three adults, if there is a choice? What if the car adapted its model in an unpredictable way 'deciding' to hit as many people as possible? You might actually end up with a Preferences setting in your car that allows you to set these choices ahead of time.

Here's another example: in a burning building, there is a chance to save either a baby or a suitcase full of £1 million. Most people's instinctive reaction is to save the baby, but if you think about it, perhaps the best action for society would be the suitcase of money, as you could probably save many more babies' lives using that money. What is the right decision for society? We're at an intriguing time in humanity whereby we are having to start to codify our basic ethics.

Liability is also a big challenge with these adaptive systems. If a CEO decides that his company should make an AI-powered machine that administers medication to people, who is liable if that machines makes a bad medication decision because it has learnt the wrong thing since it left the factory? It's almost impossible to predict how these algorithms will behave. Just look at the Flash Crash that hit the financial market in 2010.

The third type of AI to consider is where the machines become more intelligent than humans—this is the Singularity. I don't believe that we can build ethics into a super-intelligent machine (this is the Control Problem)—they will have to be learnt. You could, perhaps, exploit game theory to help it learn some ethics. Most people know of the Prisoner's Dilemma—the winning strategy is always one of 'tit for tat', that is, follow what your opponent did previously. This is the sort of ethical approach that a super-intelligent machine could learn, that has been embedded into humanity over thousands of years: "do unto others as you would have them do unto you."

Obviously, the big problem with super-intelligent machines is that they may not be concerned with the impact on humans for whatever plans they have, and this is perhaps one of our greatest existential risks. One thought I've had is that we should perhaps help them decide to depart our planet and leave us in peace, a bit like how we might have made an offering to the gods in ancient times in the hope they'd have mercy on us.

AB: That's a lot of different ethical issues to consider, especially if the future of the planet is at stake.

DH: We may not see the Singularity come along anytime soon but we absolutely have to think about the impact of an AI-driven capitalistic society, and to work out—both individually and collectively—how we can create a radically better society today.

9

Industrialising AI

Introduction

This book is predominately about how to introduce AI into your business, how to take those first steps in understanding and then implementing prototype AI capabilities. But at some point, you are going to want to move beyond that, especially if you think that AI has the potential to transform your business.

So far you should have a good grasp of what the different AI capabilities are, where they are being used effectively in other businesses, the best way to create an AI strategy and start building AI prototypes, and being aware of some of the risks involved.

This chapter then is about how to industrialise those AI learnings and capabilities within your business. It focuses on how you move from a few projects to a mature capability that can implement most, if not all, of the AI requirements that your business will need in the future.

The AI strategy that was discussed in some detail in Chap. 6 will have given you an excellent starting point. It should include your AI Ambitions which will determine how far down the industrialisation road you want and need to go. I have assumed in this chapter that you will want to at least create a permanent AI capability within your organisation, one that is able to be a focal point for AI activity, and act as a catalyst for ideas and opportunities. If your ambitions are more or less than this, then just dial up or down my recommendations appropriately.

You will also recall from the Change Management section in Chap. 6 that there is often a 'slump' after the initial success of a prototype. Now is the time to 'go large' and ensure that the momentum from those early wins is translated

© The Author(s) 2018
A. Burgess, *The Executive Guide to Artificial Intelligence*,
https://doi.org/10.1007/978-3-319-63820-1_9

into continued success. Having a robust plan for how that will happen will go a long way to achieving that. This chapter will give you the foundations for that plan.

Building an AI Eco-System

You may already have been working with a number of third parties to help you build a strategy and the initial AI project builds. Some of them you may want to continue working with on a longer-term basis; others may have just been useful for a specific task and are no longer required. If you are going to build a long-term AI capability though, you will need to start thinking about how much of it will be supported by third parties, and which pieces you intend to build up internally.

A useful way to think of this is as an 'Automation Eco-System'. This may include software vendors, strategists, implementation partners, change management experts and support teams, some of whom will be internal and some of whom will be external. Some may have a bigger part to play in the early days, whilst others will become more important once some of the projects have been firmly established.

If you are planning to use a third party to provide any of the Eco-System capabilities, be aware that some may be able to provide more than one; an implementation partner, for example, may also have change management capabilities. Whether you use both those capabilities from that provider will, of course, be up to you. I cover some of the subtleties of this in the section on Vendor Selection later in this chapter.

An Automation Eco-System may end up looking something like this (Fig. 9.1):

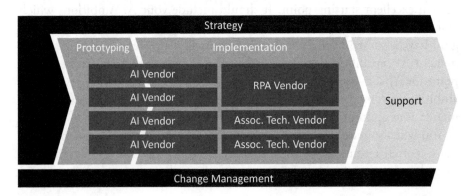

Fig. 9.1 AI Eco-System

Each of the capabilities can be provided internally or externally. Apart from the software vendors, the roles can cover AI specifically, or Automation (including, say, RPA) generally. I will cover each of these roles in turn.

Strategy This area concerns the sorts of things you are reading about in this book, particularly those in Chap. 6. These are the activities that I, as an automation advisor, carry out for my clients. It involves building the Automation Strategy, which will include the Maturity Matrix, the Heat Map, the Business Case and the Roadmap. It will likely involve the technical approach (vendor versus platform versus bespoke) and, connected to that, the approach to this Eco-System. It can also include supporting the selection of the software vendors and the other service providers. Most importantly, it should provide thought leadership to the organisation.

As the AI market is very complex but relatively young there are very few advisors that are able to cover the business strategy aspects as well as the technical approaches. Some of these (including myself) are able to provide prototyping services as well. The connection between these two capabilities is pretty tight; therefore, it may make sense to cover both of these with one provider.

Although the bulk of the strategy work is carried out up-front, it can be beneficial to retain the strategy advisor longer term to ensure that the benefits are realised and the internal capability can be realised. Most clients who wish to pursue a predominantly internal capability (with little dependency on external parties) keep a light-touch external strategic involvement to provide appropriate checks and balances.

Prototyping The building of the prototypes, pilots and PoCs requires a flexible, agile approach, and follows on closely from the output of the automation strategy (and, in some cases, can form part of the automation strategy). There are not (at the time of writing) many consultancies specifically focused on prototyping—this capability is generally provided by either a few of the strategy advisors, or by the implementation providers.

The activities are those that are largely covered in Chap. 7 of this book, including the build of PoCs, RATs, MVPs and Pilots. If your company is just starting its AI journey then it makes sense to buy in this capability until you have enough experienced and skilled resources in-house.

AI Vendor The AI Vendor could include a packaged software provider, a platform provider or a vendor of some of the various development tools that are used to create bespoke AI builds, depending on the technical approach being taken.

For some of the platform providers, and maybe some of the development tools, you may have existing relationships in place. No matter what arrangement though, the software obviously sits at the core of your Automation Eco-System, and many of the other members of that eco-system will be dependent on those choices.

It should also be noted that some of the packaged software vendors can provide their own implementation resources. This is usually the case for those that have yet developed their software to be 'implementation partner ready'. One or two of the platform providers, such as IBM, also have large implementation service capabilities, although other third-party providers can also be used with their platforms.

It will be likely that you will end up with a number of different software vendor relationships, irrespective of which technical approach you take. Even with a platform-based strategy, it is probable that you will need to bring in other vendors to supplement any gaps in the platform capabilities.

Implementation The Implementation Partners are usually brought in following successful prototyping (unless they are carrying out those activities as well). They are generally concerned with the industrialisation of the AI capability and can provide a full range of development resources. The Big Four global consultancies can provide these sorts of services, as well as a number of their smaller competitors.

Whether you need an Implementation Partner at all will depend on how keen you are to build an internal capability, and how quickly you want to get there. A Strategy Advisor and a Prototyping Partner can get you a long way towards self-sufficiency, but if you want to move fast (or if you are not interested in building an internal capability at all) and have deep pockets then an Implementation Partner makes sense.

An Implementation Partner will be able to continue the work started during the Strategy and Prototyping stages, and help build and 'bed in' those applications. Depending on their capabilities they may be able to build further applications as well as deliver some or all the Change Management requirements.

RPA and Associated Technologies Vendor As I detailed in Chap. 4, AI is rarely the only technology you will need to deploy to realise the full benefits available. You may also need to bring in an RPA vendor, a Cloud vendor, a Workflow vendor, and a Crowd Sourcing vendor, depending on the complexity of your solution and what existing relationships you already have (many enterprises looking at AI will already have a Cloud capability, for example).

Some vendors will work better together, either technically or culturally, so do consider this if you have a choice. Build these eco-system members around your AI vendors, looking at who they have worked well with before and whether there may be any particular integration challenges or not.

Your RPA vendor, should you require one, will be an important choice. Some AI vendors have strategic relationships with RPA vendors, and some Implementation Partners may also have 'preferred' RPA vendors that they work with. In a very few examples (to date) there are software vendors that can provide both RPA and AI capabilities, with some of these also offering Crowd Sourcing as well. Generally though, most enterprises bring in separate vendors for these crucial capabilities.

Change Management This is an important capability for your eco-system and should be considered early on in the lifecycle. Many large organisations will be able to provide a general change management capability internally, but the specific characteristics of an AI project should be borne in mind to determine whether this will be the most appropriate approach.

The capability can be provided by a generalist third-party provider but those that can claim to have experience in the unique characteristics of an AI implementation will usually be preferable (the exceptions being when you have an existing, trusted provider or you need one with experience in your specific industry sector).

Support With this I mean the ongoing management of the developed AI solutions. That could mean ensuring that the applications are available through management of the underlying infrastructure and networks, or ensuring that the application is providing accurate and meaningful results. It could also mean the debugging of any bespoke or procured software. Each of these activities will generally be done by different parties and will depend heavily on the technical approach that you take, any existing IT policies and the complexity of the developed solutions.

For infrastructure support, you may have passed the responsibility to an outsourcing provider or a Cloud provider, but if you have decided to keep it in-house your IT team may have to support some unfamiliar technologies, including the management of high-powered GPUs and large storage arrays.

Management of the AI applications (apart from actual code errors) could be carried out in-house if you have built up the necessary capability, probably in an Automation Centre of Excellence, which I describe later in this chapter. It may well involve recruiting AI developers and data scientists, neither of which are readily available or cheap at the moment. There are very few providers, though, that specialise in supporting AI applications—they tend to be either the Prototyping firms or the Implementation providers. The skills will need to include the ability to understand the business requirements, the data and the technology, hence the challenge in finding the perfect provider, if that is the route you are taking.

If the procured or bespoke-build software is faulty, then it will usually be down to the software vendor to fix it—this is the classic 'third line' support and should be included in any contracts between your organisation and the vendor.

I have not included Sourcing Advisors in this eco-system. They are likely to be a short-term requirement at the start of the journey, and therefore would not form part of an on-going capability. I do talk more about the role of Sourcing Advisors in the next section though.

Whether you end up creating an eco-system with these elements, or you break it into smaller chunks, it is important to consider all the aspects I have mentioned above. For those elements that you plan to buy-in, you will need to have a robust selection process, which I cover in the next section.

Selecting the Right AI Vendor

To a large extent, selecting AI software vendors and service providers will follow the same rules as with any IT vendor or provider, but there are some key differences with AI that can present some additional challenges as well as opportunities.

One of the first questions that will need to be answered is what exactly is being procured. For software, the answer to this will depend on the technical approach that has been decided upon—off-the-shelf, platform or bespoke build (or any combination of these). There is likely to be one piece of software that is at the centre of most things, particularly in the early stages, and this should be the initial focus. This could be the off-the-shelf software capability that is required for the pilot, or the platform that the bulk of the applications will be built out from.

Your organisation may be constrained in what software it can choose. This may be due to the IT strategy stipulating particular criteria (e.g. it must operate as-a-service), specific standards it wants to adhere to (e.g. no JavaScript) or general procurement rules (e.g. third-party company must be at least two years old). Beyond these constraints though, it is important to consider the following when selecting AI software vendors:

- **Proof of capability**—because there is so much hype in the market many vendors will exaggerate their AI credentials in order to attract attention to themselves. Some will have only a minimal AI capability as part of their overall solution (maybe a bit of simple NLU embedded in it somewhere) but claim it all to be 'powered by AI'.

  Now, of course, the solution you actually require may not necessarily need full-on AI, but if you have a specific application in mind then you will need to make sure you can cut through the hype. The understanding of AI capabilities that you will have gained from this book will hopefully go a long way to helping achieve that, but you can also call on external advice as well if need be.

  An important step in being able to understand and prove the capability of the vendor will be to have a comprehensive demonstration of the system and, if appropriate, a small test. Tests, as with pilots and PoCs, can be tricky for AI solutions as they generally need a lot of data and training to get them to work effectively. If the vendor can do something meaningful here though, do take up the offer.

- **Proof of value**—you should already have at least an outline business case by this stage, but it is useful to understand how the vendor measures the value of their solution. Some will have an approach that allows you to quickly assess their value whilst others may expect you to do all the hard work on this (this may, of course, mean that they will struggle to create any value from their solution, and it should be seen as a potential warning). If the vendor does have a model that can validate your own business case assumptions, then do take them up on that approach. The key thing is for the value that they can demonstrate to align as much as possible with your own business case.

- **References**—taking up client references should be an obvious thing to do, but many enterprises don't bother (and live to regret it later). The challenge with an immature market like AI means that client references and case studies will be thinner on the ground. How far you want to pursue this will depend on the perceived fit of the vendor's solution to your requirements, their uniqueness and your appetite for risk. If you want to bring in a vendor who has no existing clients, then you may want to see what opportunities there are for you to be a 'foundation client' for them— this is covered below.

- **Pricing model**—the commercial aspects are obviously very important when selecting a vendor, but it's not all about the headline price. In Chap. 6 I described some of the different pricing models that are being used, and you should explore which ones might be the most relevant to you, and whether the vendor can provide them. For variable pricing, model different scenarios, especially if it materially impacts the business case—a price per API call may sound attractive at first, but if your application is very popular, those escalating costs may start to outweigh the benefits. Also, seriously consider gain-share or risk-reward models, but make sure that they are based on real numbers that you can measure easily.

- **Cultural fit**—although you are just buying some software, it is always worth finding a vendor company that has a good cultural fit with yours. The AI project may be a long, and sometimes stressful, journey for both parties, so you need to know that there is some common ground and purpose between both of you. Cultural fit can be assessed simply from having a good feeling about it, but it can also be worth trying to do this more objectively, especially if you want it to form part of your selection criteria.

- **Future proofing**—ensuring that the software you buy will still be relevant and working in five years' time can be difficult to determine, but it should definitely be considered carefully. In the previous chapter, I talked about the risks of selecting the wrong technology and those considerations can be applied here. But it is probably the more traditional elements of the technology that are the ones that are most at risk of going out of date. Ask what technologies the software is built on, and what technology standards does it rely on. External expert advice may also be useful here.

- **Foundation client**—procuring software that has not been used in anger with any other clients may seem like an unnecessary risk, but there can be benefits from taking this approach. All other things being equal (you have tested their capability, the technology is good and future-proofed and there is a good cultural fit) then, versus a more established vendor, a new vendor may be willing to offer you some significant benefits in return for taking the risk of being their first client. Usually the benefits amount to heavy discounts, especially if you also agree to promote your use of the software. But it can also mean that you get the opportunity to shape the development of the solution, and make it more closely match your own requirements. This will also ensure that you get the undivided attention of the vendor's best people, and that can count for an awful lot in delivering a successful implementation.

- **Technical requirements**—it should go without saying that the technology needs to fit into your current environment and align with your IT standards and strategy. But with AI you will also need to consider the data. Is your data the right sort of data for the vendor's solution, is it of good enough quality and quantity? These are key questions that must be satisfied early on in the selection process.
- **Professional services requirements**—the final consideration is how the vendor's software will be implemented. What professional services are required, and who can provide them (the vendor, third parties or your own organisation)? If it requires a third party, then understand how many providers out there can do it, and which ones are most familiar with the specific vendor's software. Some vendors may have a partnership program where they will have certified or approved service providers. If you are already developing your eco-system, then you will need to determine how much commonality there is between these. You will also obviously need to evaluate the cost of the services at this stage. The selection of service providers is covered in the next section.

Selecting the Right AI Service Provider

In the earlier section concerning the Automation Eco-System I also talked about the type of service providers that may be required, including strategy, change management and implementation consultancies—these will also have their unique AI-related challenges and opportunities when procuring them in.

Each of the services capabilities you require should be considered as separate 'work packages', which can be procured individually or in groups. From the Eco-System considerations, you should have a starting point for which capabilities need to be bought in, and how they might be grouped together. For example, you may think that you need to buy-in Strategic Advisory, Implementation Services and Change Management, but you think that you will be able to handle the Support Services in-house, and you also think that a single provider could do Implementation and Change Management. In this example then you would be looking for two providers: one for the Strategic Advisory work package and one to handle the Implementation and Change Management work packages.

During the selection process (or processes) these initial assumptions should be tested, but they do provide a useful starting point. You should therefore provide the flexibility in your Request for Proposal (or whichever procurement method you are using) for providers to bid for each work package separately and combined. You may find as you go through the procurement process that one provider is really good at Implementation, while another is better at Change Management—in this case you can always adjust your approach to instead procure two separate providers for these work packages.

You should also test your in-house capability (if you have one) against the external providers if you are not sure whether to 'buy or build'. You can treat them as a bidder in the procurement process and even price up the cost of their services.

Another important consideration when looking to identify potential providers is your incumbent or strategic providers. Many large organisations have preferred suppliers and you may have to align with these. AI is such a specialised technology, though, that (unless your incumbents are actually AI specialists) you should argue strongly to bring in providers with the necessary skills and experience in AI.

The strategic advisory role should be your first procurement consideration, as many of the other decisions will be based on the Automation Strategy that is developed with your advisor. Robust, demonstrable methodologies, independence and (maybe most crucially) cultural fit will likely be strong factors in your selection here. For the procurement of the remaining work packages, I think the following are the key AI-specific considerations, whether the resources come from an established provider, software vendors or contractors (I have used the words 'service provider' below to cover these):

- **Experience**—it may be difficult for a lot of providers to demonstrate a real depth of experience in implementing AI, simply because the market is so young, so you will need to bear this in mind. You will need to look for capability in the particular approaches and tools that you have decided upon and any relevant partnerships with software vendors (see below). There will be some experience that is technically specific (skills in a certain tool, for example) and others that is more industry sector specific (e.g. ability to understand customer data for telecoms firms), and there will realistically be a compromise that has to be made between these two (unless you are lucky enough to find the perfect match). It may be that you have the data knowledge already in your organisation.
- **Partnerships and Independence**—for the tools and approaches that you have selected, you will want to look for a service provider that has a strong relationship with these. Many vendors have 'partnerships' with providers, which can range from a loose connection to an almost symbiotic one—it is important to understand the level and history of those partnerships.

In most cases the service provider will be rewarded by the vendor for selling their software licences. But you may also want your service provider to be independent of the software vendors, and a balance has to be struck here. Probably the ideal arrangement is where the service provider has strong, independent relationships with a number of vendors and will be happy to recommend and implement any of them. If you can't find true independence in the service providers then you will need to rely on your strategic advisor to provide this.

- **Cultural Fit**—this is often overlooked in a service provider selection process but, in my experience, is the one factor that causes most relationships to break down. Because your AI projects could be quite fraught at times, you will want to work with a provider that shares your values and objectives, and has people that get on with your people. Cultural fit can be formally measured during the selection process or it can be assessed informally, but it should not be ignored.

- **Approach**—most service providers will have methodologies, and these should, of course, be evaluated and compared carefully. But you should also try and look 'under the bonnet' (or 'under the hood') to see how they may approach different challenges, such as testing, poor data or biased data. Their answers should tell you a lot about their underlying AI capabilities.

- **Pricing Model**—as with the software vendors, pricing is going to be important but so is the pricing model. With service providers, there is the opportunity to introduce 'gain share' or 'risk/reward'-type approaches where the provider will share some of the risks but also some of the gains of the project. This helps align their objectives with yours. The challenge with these pricing models is getting practical metrics—they need to be general enough that they relate to business outcomes but specific enough that they can be measured. The metrics also need to be timely—service providers will not agree to a target that can be measured only in two years' time, for example. Your strategic advisor, as an independent resource, should be able to help you identify and implement practical KPIs.

- **Knowledge Transfer**—the final consideration is how the service provider will transfer all the relevant knowledge to your internal team. This assumes that, at some point in the future, you will not want to be dependent on a third party for your AI capability—some enterprises want to get to this point as quickly as possible, whilst others are happy to remain dependent for as long as it takes. The knowledge that will be transferred from the service provider to your teams will include a range of things, and will depend on the relationship and contract you have with them. It may involve some licensing of Intellectual Property (IP) from the provider (such as methodologies and tools), but will generally be in the form of work shadowing and the hands-on experience gained by your team. As your resources get more skilled, and you bring new employee resources in, the service provider can start to back away; as you ramp up, they will ramp down.

There is another potential third party that you may wish to engage in. Sourcing Advisors are experienced procurement specialists that can help you with both the vendor and service provider selection processes. They will be able to help create a sourcing strategy, find appropriate companies to evaluate, build and manage the selection process, and support negotiations.

The over-riding requirement for any sourcing advisor is their independence. Then you should look for capability in AI and in your specific industry sector. If they do not understand the AI market well enough (it is, after all, sprawling and dynamic) then you may want to team them up with your strategic advisor (or find someone that can do both).

With both the vendor and the service provider selection there will need to be a fine balance between a robust selection process and the flexibility that dealing in a young, dynamic technology requires. Be prepared to change your plans if things don't initially work out, or better options come along. As the cliché goes, be ready to 'fail fast'. And this means ensuring that the arrangements you have in place with your vendors and providers are as flexible as possible, whilst still demonstrating your commitment to them. Work closely with your procurement department to make sure that they understand the special nature of AI projects and the demands that it may place on their usual methodologies.

Now that we have an approach for bringing in the necessary third parties, we need to look at the internal organisation and the skills and people that might be needed.

Building an AI Organisation

There are a number of things to consider when building an organisational capability to manage your AI (and other automation) efforts, all of which will be based on how bold your AI ambitions are. You will remember from Chap. 6, as part of building an Automation Strategy, I stressed how important it was to understand your ultimate AI ambitions: do you want to simply 'tick the AI box', to improve some processes, to transform your function or business, or even create new businesses and service lines? Assuming that you are somewhere in the middle of this range, that is, you want to improve processes and transform some areas of your business through AI, then you will need to create a team or even a CoE to ensure that you are successful and can extract the maximum value from your efforts.

In this section I will describe what an AI CoE could look like. Based on the skills, capabilities and ambitions of your own organisation you should be able to assess which elements are relevant for you, and therefore what your own CoE might look like. I will also describe how a CoE can be integrated into an overall company organisation.

It is important for an Automation CoE to have a mission. This will describe, and help others understand, its purpose. You will be able to find the words that are most relevant to your own requirements and objectives of your company, but generally the CoE's mission should be focused on driving the introduction and adoption of AI technologies. It should act as a central control point for assessing AI technologies and monitoring progress of ongoing projects. Perhaps most importantly, it should provide leadership, best practices and support for projects and teams implementing AI solutions within the business.

Two of the key inputs to determine the scale and structure of your CoE, apart from your AI ambition, are your AI Heat Map and Roadmap. These will list the functions, services and processes that are being targeted for automation, and will give the priority of how they are likely to be rolled out. Particular aspects to consider will be the different technologies that will need to be deployed, the complexity of the solutions and the current state of the data that will be exploited.

The roles that are generally included within an Automation (and particularly an AI) CoE fall into four main functions: the Management of the CoE, an overall Architecture Management, the Implementation Teams and Operations. Some of these roles, including, particularly, parts of the Implementation Teams, can be provided by third-party providers or vendors, especially in the early days of the CoE. The following is a guide for how some organisations are structuring their automation CoEs, but can obviously be adjusted for your own needs and aligned with your company culture.

The **Management Team** should ideally start out relatively small, including at least a manager and some project management capability. The manager should have responsibility for the whole CoE and the people assigned to it, as well as the communications across the organisation and upwards (communications could be taken by an internal communications specialist later on). For management and control of projects, the team (which could just be one person at first) should be responsible for planning, project management, resource tracking and reporting across all the various AI projects in the CoE. Other key areas that usually come under this team would include the co-ordination of training and education for both members of the CoE and users of the systems.

Taking on the formulation of the business case and solutions are the **Architect Team**. This team is best headed up by someone who understands the business functions and processes but also the technical aspects as well. The team will look at the opportunities for automation in the wider context (so would have control of the AI Heat Map) and would create business cases for each of them, carrying out the initial scoping of effort and technologies, all of which would be done in close collaboration with the Subject Matter Experts (or Process Owners or Business Analysts) from the Implementation Team.

The Architect Team would be responsible for managing the pipeline of opportunities, ensuring that these are being proactively sought, and reactively received, from across the business. They would also include technical architects (unless this role is retained within the IT Department), data scientists (if required) as well as managing customer experience.

The **Implementation Teams** are where the bulk of the CoE resources would reside. It would consist of a number of project teams, each focused on a specific solution. Depending on the size and complexity of each project they could include: a project manager, a project architect, developers (who would have responsibility for importing and training data, creating and configuring the models), SMEs/Business Analysts/Process Owners (carrying out solution design and validation) and Quality Assurance. If appropriate, specialist resources such as customer experience specialists, linguists, web developers and integration developers could also be assigned to projects.

Agile development approaches are the most suited to building AI solutions. The specific type of Agile (e.g. Kanban, Scrum) is not that important, just that the developers and SMEs work closely together and on fast iterations. Because of this, and the transformational nature of AI, the members of the Implementation Team (and the CoE generally) should have a strong mix of both business and technical knowledge. For regulated processes, which throw up additional control requirements, there are a number of Agile approaches that try to cater for these, including R-Scrum and SafeScrum.

Support and maintenance of the released applications (fixing bugs and carrying out any enhancements) can be handled in a couple of different ways. For small implementations, the original project team can retain responsibility for support, whilst for larger or more complex projects a separate support team should be created. Therefore, young or small CoEs have the support responsibility within the Implementation Team, whereas larger, more mature

CoEs will generally create a new team for this purpose, or have it embedded as part of the Operations Team.

The final group is the **Operations Team**. This team will have responsibility for the deployment, testing, updating and upgrading of any of the live systems. They will also have responsibility for the technical integration of the AI solutions with any other systems. For organisations used to taking a DevOps approach (where the Operations and Development resources work as integrated teams), then much of this Operations Team would be subsumed into the Implementation Team.

As I mentioned earlier, the above groups are only meant as a guide and can be flexed to match your own company's requirements and practices. Another consideration regarding the setting up of a CoE is how it fits into the overall organisation structure of the company.

Some companies, such as Aviva and Clydesdale and Yorkshire Bank, have created 'lab' or 'innovation hub' environments (the Aviva one is actually called a Digital Garage). These are useful vehicles to generate additional buzz around the initiatives and to foster a culture of innovation. They usually focus on the early stage development of ideas. CoEs though tend to be a little more practical and will include most, if not all, of the capabilities to identify, build and run new technology initiatives.

A key consideration, especially for larger organisations, is whether to have a single, centralised CoE or to spread the capability across the various divisions or business units whilst retaining a central control function. The approach that usually works best for AI, at least initially, is to keep everything as centralised as possible. This is for a couple of reasons.

Firstly, the impetus and momentum for the early AI initiatives usually come from one area of the business (because of greatest need, biggest opportunities, most enthusiastic management, etc.), and this can form the genesis of the CoE. When other business units see the benefits that are being generated, then they can tap into the existing capability rather than creating it themselves from scratch.

Secondly, an 'independent' team not involved in the day-to-day running of the function is much more likely to identify, and be able to implement, transformational change. Left to their own devices, business units rarely consider the full range of opportunities available and will instead focus on the simpler process improvements. Both are valid types of opportunities, but AI has much to offer with regard to transformational change and therefore it should be promoted at every opportunity.

A final consideration regarding the organisational structure of your industrialised AI capability is whether to appoint a senior executive to oversee it all.

Many large companies, especially in the financial services sector, already have Chief Data Officers (CDO). These roles take responsibility for the enterprise-wide governance and utilisation of information as an asset (including in some cases a large element of revenue generation). For some companies, such as Chartis, Allstate and Fidelity, the CDO has a big influence on the overall strategy of the company.

A relatively new position that is starting to be seen in some businesses is the Chief Automation Officer (CAO). These can sometimes be called Chief Robotics Officers or Chief AI Officers (CAIO). This role should try to embed automation, and AI, at the centre of the business strategy. A CAO, more than a CDO or a CTO, can look outward and forward. They will tend to have a greater focus on the business opportunities of automation than, say, a CIO. (Some analysts believe that a CAO is more suited to a move up to CEO than a CIO might be.)

CAOs are still relatively rare, and it may be that it is a role that is needed for a certain period of time, to get automation up and running and firmly embedded in an organisation, and then have those responsibilities eventually absorbed back into business-as-usual.

I am not aware, at time of writing, of any organisations with CAIOs yet, although there is plenty of talk about the role. Again, it might be a short-lived necessity, and many people are sceptical that this level of focus is required in the boardroom. It may be that the CDO or CAO (if they exist in a company) are able to cover AI specifically.

Whether an organisation chooses to bring in a CDO, CAO or CAIO, the people carrying them out need to have some pretty special capabilities: they will need at least a reasonable understanding of the technologies and the data infrastructure; they will need to be able to work cross-functionally, since auto-mation opportunities can exist across the whole business and will require the cooperation of a number of departments; they will need to behave like entrepreneurs within the business and they will need to have the industry standing and inter-personal skills to attract and retain the best talent. That is not an insubstantial shopping list.

We have now come to end of the journey to understand, create and indus-trialise AI in businesses. Throughout the book, I have focused specifically on what AI can do for organisations today, and the practical approaches that can be taken to exploit the inherent value in this technology. But now, in the final chapter, it is time to look to the future: how AI may develop, what new opportunities and risks it will create, and how we can best plan to get the most from it.

The Data Scientist's View

This is an extract from a dialogue with Richard Benjamins, who, at the time of the interview, was working as Director, External Positioning & Big Data for Social Good at LUCA, part of Telefonica, a European Telecoms Company. He is now Group Chief Data Officer & Head of the Data Innovation Lab, AXA, a global insurance company.

AB: In your role at Telefonica's data-led business, LUCA, you understand the inherent value, as well as the challenges, in using Big Data for AI applications. Can you tell me more about this?

RB: Firstly, you have to look at the numbers. Six years ago, McKinsey estimated that Big Data would bring $300bn in value for healthcare and €250bn for the European Public Sector, but in their most recent update at the end of 2016, they reckon that the numbers were actually 30% higher than this.

But none of that tells you how you should get the value from your own big data and how to measure it, and that's the biggest challenge for organisations who want to be able to realise that value. I think there are four ways that you can get at that inherent value.

Firstly, and probably most simply, you can reduce the cost of the IT infrastructure that is being used to manage your big data, using open-source tools such as Hadoop. This can save significant amounts of money, and is easy to measure. Secondly, big data can be used to improve the efficiencies of your business, that is, enable you to do more with less. Thirdly, it can be used to generate new revenue streams from your existing business—this can be a challenge though as it is difficult to know where to start, and it can be hard to measure the value. The final source of value is where completely new revenue streams are created from big data, that is, new products where data is at its centre, or by creating insights from data to help other organisations optimise their own businesses.

I believe that in the next few years, the dominant source of big data value will be from business optimisation, that is, by transforming companies into data-driven organisations that take data-driven decisions.

AB: So how do you start to put a number against business optimisation through big data?

RB: This is the big challenge. Of course, you should have a base case and compare the difference before and after, but big data initiatives are rarely the only reason for doing something, so it's difficult to isolate the value. There are two ways to help with this though. The first is to do a segmented experiment; for example, have one group of customers that are being analysed through big data, and another that are not (and maybe even another where a different approach is being used). The second is to do something with big data that has never been done before – that way you can compare the results directly with your base case.

From my experience, though, mature organisations know that there is value in big data and will not be overly obsessed with trying to measure every bit of value. When companies reach the stage where big data has become business-as-usual, then it changes the way that departments interact and the value naturally flows from this.

AB: In order to manage all of this data and value, the role of the Chief Data Officer (CDO) must be crucial?

RB: Definitely. We are seeing a big increase in the number of businesses that have a CDO. According to one survey I have seen, 54% of companies now have a CDO, up from just 12% in 2012.

Where the CDO sits in the organisation is an open question still. It's getting closer to the CEO level every year—in Telefonica for example, five years ago the CDO was originally five levels below CEO but now reports directly into that position.

The best position for a CDO to be in is where they have responsibilities across various functions or departments, so that their role, and objectives, are not defined by a single department. Reporting into the COO is probably a good balance as they will then have good cross-functional visibility.

AB: Big Data is obviously closely related to Artificial Intelligence. Do people get confused between these terms?

RB: The term Artificial Intelligence can be used for many things including Big Data and Machine Learning. The hype that is around AI can be useful because it brings an increase in interest and attention but it is important to keep in mind what we are talking about. It is not just about building wonderful applications, it is also about asking fundamental questions related to how people think, how they solve problems and how they approach new situations. When thinking about AI I think it is important to consider these three things: AI can solve complex problems which used to be performed by people only; what we consider today as AI may just become commodity software; and, AI may help us shed light on how humans think and solve problems.

And remember, in the big scheme of things, this is only the beginning...

10

Where Next for AI?

Introduction

This book has always been about what executives need to do today to start their AI journey. Rightly so, but it is also important to understand what is coming down the line. Organisations need to be prepared as much as possible to exploit any future advancements as well as mitigate any of the developing risks.

This final chapter tries to predict the near future for AI. This is, of course, a bit of guessing game, but hopefully it is a well-informed one. Firstly, I look at which of the AI capabilities I have discussed in the book are likely to flourish and thrive the most, in other words which should be watched most closely for significant developments.

I then have a go at predicting when AI might become 'business as usual', that is, when we are no longer talking about it as a new and exciting thing, in a similar way that outsourcing has now become simply another tool for the executive to deploy. This section covers some of the more general predictions about AI and its use in business.

And finally, I close the book with some words of advice about how to future-proof your business (hint: it includes AI).

© The Author(s) 2018
A. Burgess, *The Executive Guide to Artificial Intelligence*,
https://doi.org/10.1007/978-3-319-63820-1_10

The Next Opportunities for AI

In Chap. 3 I described the core capabilities of AI and then in Chap. 5 I gave numerous examples of how those capabilities are being used in businesses today. But the technology will inevitably get better, and the speed, accuracy and value of these capabilities will only increase, probably at an ever-accelerating pace.

This section provides my summary of how each of those capabilities might develop over the next few years—some will accelerate whilst others may hit some bumps in the road.

- **Image recognition** has already advanced dramatically over the last five years and, particularly, over the last couple. Its development over the next few years will continue at this heady pace as more image sets become available for training purposes, and better algorithms, faster processors and ubiquitous storage will all contribute to the ability to improve accuracy but also tag moving images as well. There has already been some progress in this area but I think that we will see objects in films, clips and YouTube movies identified and tagged automatically and efficiently. This would mean you could, for example, search for an image of a DeLorean car across all of the movies online, and it would return the frames in the films where the car appeared.

  The next development beyond that will be to identify faces in movies, just as Google Images and Apple's Photo applications can do today for still photos. Being an optimist I am hoping that the medical applications of image recognition, such as in radiology, will really start to take off—they will have a huge impact on society, and therefore deserve to be promoted.

  The technique of having AI systems play (or battle) with each other, called GANs, will enable systems to create new images from scratch. Put simply, one system, after an initial period of training, will create an image, and another will try and work out what it is. The first will keep adjusting the image until the second system gets the right answer. This technique could also be used to create scenery for video games or de-blur video images that are too pixelated.

  Image recognition will have to face the challenges around bias, which are certainly coming to the fore at the time of writing. Ensuring unbiased image sets should be a mission for all those creating and controlling data, and it shouldn't become an issue only when things have clearly gone wrong.

- **Speech Recognition**, as can be evidenced in domestic devices such as Amazon's Alexa, Google's Home and Apple's HomePod, is already very capable, although there are plenty of examples of poor implementations as well. I predict that this capability will become much more prevalent in a business-to-business environment, powering, for example, an HR Helpdesk, or a room booking system.

  Real-time voice transcription already works reasonably well in a controlled environment (a nice, quiet office with a microphone on a headset) but advances in microphone technology in parallel with better algorithms will make speech recognition viable in most environments.

  In the section on malicious acts in Chap. 8 I talked about voice cloning. As long as this technology can stay out of trouble (or the people wanting to use it for good can keep one step ahead of the hackers) then this has the potential to help many people who have voice-crippling diseases.

- **Search** capability really took off in 2017, especially in specialised sectors such as legal. There are a number of very capable software vendors in the market which means that this will be a useful introduction for businesses as they look to adopt some level of AI capability. The algorithms will inevitably get better and more accurate, but I think that the future for this capability is as part of a wider solution, either by extending its own capabilities (e.g. to be able to auto-answer emails from customers) or by embedding it in others (as part of an insurance claims processing system, for example).

- **Natural language understanding** is closely connected to both Speech Recognition and Search, and will benefit from the advances there, particularly in areas such as real-time language translation. But I think the hype and elevated expectations that have surrounded chatbots will have a detrimental effect on their adoption. I predict that there will be a backlash against chatbots before they start to make a recovery and become widely adopted and used every day. Those firms that implement chatbots properly (i.e. for the right use cases, with appropriate training, and capable technology) will see successes, but those that try and take shortcuts will speed the oncoming backlash.

  The sub-set of NLP that focuses on generating natural language (NLG) will see some big strides, and therefore wider adoption, over the next few years. It will benefit from broader acceptance of AI-generated reports, particularly where they add real value to people's lives, such as with personalised, hyper-local weather forecasts. As more data is made available publicly, then more ideas will surface on how to extract value from it and to communicate those insights to us using natural language.

- **Clustering** and **prediction** capabilities can, for this purpose, be considered together. Two things will happen regarding the use of big data: firstly, as businesses become more 'data aware', they will start to make better use of the data they have, and also exploit publicly available data as well; and, secondly, less data will be needed to get decent results from AI, as algorithms become more efficient and data will become 'cleaner' and (hopefully) less biased.

  Financial Services will be the main beneficiary from Prediction, being used in more and more companies (not just the global banks) to combat fraud and up-sell products and services. Cross- and up-selling will also become the norm for retailers.

  The challenge for these capabilities will be whether they will be used productively and ethically. Increasing the number of click-throughs and delivering targeted adverts may make sense for some businesses, but it is never going to change the world, and will no doubt frustrate many customers. And, of course, the questions around bias and naivety will remain on the table and will need to be addressed for every use case.

- **Optimisation**, out of all the capabilities, has, for me, the greatest promise. The work that is being done with Reinforcement Learning (that was at the heart of AlphaGo's victory) has enormous potential in the worlds of risk modelling and process modelling, and specific examples where energy bills can be significantly reduced (as at Google's data centre) demonstrate that it should have broad and important applications for us all.

  One of the biggest constraints on AI right now is the availability of properly labelled, high-quality data sets. GANs, which I mentioned earlier, have the potential to allow AI systems to learn with unlabelled data. They would, in effect, make assumptions about unknown data, which can then be checked (or, more accurately, challenged) by the adversarial system. They could, for example, create 'fake but realistic' medical records that could then be used to train other AI systems, which would also avoid the tricky problem of patient confidentiality.

  But reinforcement learning and GANs are the AI technologies that could still be described as 'experimental', so we may not see these being used on a day-to-day basis for a few years yet. In the meantime, we have systems that can make important decisions on our behalf, including whether we should be short-listed for a job, or approved for a loan. The main advances here, I think, will be in how they are used, rather than any major advances in the underlying technologies. Eliminating bias and providing some level of transparency will be the key areas to focus on.

• As we are looking into the future, it is worth considering some of the questions around **Understanding**, which I stated in Chap. 3 was still, as an AI capability, firmly based in the AI labs, and may be there for a very long time, if not forever. Having said that, a really interesting area of development to keep an eye on is that of Continual Learning. This is an approach, currently being developed by DeepMind, that hopes to avoid 'Catastrophic Forgetting', the inherent flaw in neural networks that means that a system designed to do one thing won't be able to learn to do another unless it completely forgets how to do the first.

Because the human brain learns incrementally it is able to apply learnings from one experience and apply it to another, but neural networks are extremely specialised and can learn only one task. DeepMind's approach, using an algorithm they call Elastic Weight Consolidation (EWC), allows the neural network to attach weight to certain connections that will be useful to learn new tasks. So far, they have demonstrated that the approach works with Atari video games, where the system was able to play a game better once it had been trained on a different game.

Although there are a number of labs looking at tackling Artificial General Intelligence, EWC is probably the most exciting opportunity we have for getting closer to an answer to this most intriguing of challenges.

When Will AI Be Business as Usual?

In the previous section I talked about the developments I expect to see in the next few years for each of the AI capabilities. In this section I want to be able to generalise those predictions and look at the wider context around AI. Specifically, I will look at the question of when (if ever) AI will become 'business as usual'.

The traditional way of understanding the maturing of any technology is to see where it is on the Gartner Hype Cycle. This is a well-used graph which plots the life of a technology from its initial 'technology trigger', through its rise in popularity to a 'peak of inflated expectations' that is then followed by a decline into the 'trough of disillusionment'. After that it recovers through the 'slope of enlightenment' and finally reaches the 'plateau of productivity'. It is this final stage that can be defined as *business-as-usual*.

The challenge with plotting AI on this curve is that AI is many different things and can be defined in many different ways. We could plot some of the individual capabilities, such as NLU (heading into the trough of disillusionment) or some of the technologies, such as machine learning (right at the peak of inflated expectations) or GANs (as a technology trigger). To generalise though, most of the machine-learning-based AI technologies are pretty much peaking with regard to expectations and are therefore primed to enter the steep slide into disillusionment.

Being in the trough doesn't mean that everyone abandons the technology because it doesn't work anymore—in fact it's usually quite the opposite. Developments will continue apace and great work will be done to create new applications and use cases. It's just that the perception of the technology across businesses and society will shift; there will be less hype, with people taking a more sober, considered approach to AI. The best thing about being in the trough of disillusionment is that everyone's expectations are now much more realistic. And that surely must be a good thing.

How long it stays in the trough will depend on a number of factors. The four key drivers I described in Chap. 2 (big data, cheaper storage, faster processors and ubiquitous connectivity) will only continue on their current trajectory. Big data may be the one area that makes or breaks it through.

I think the greater availability of public data will be one of the main reasons that AI is adopted by the mainstream and reaches the plateau of productivity quicker. But there are challenges to this, as the tech giants hoard their own proprietary data sets farmed from their billions of 'customers' (who, by the way, should more accurately be called suppliers, since they provide all the data that is used by the tech companies to sell advertising space to the advertisers, their actual customers).

But even their data will become more difficult to farm—there is a complicit deal between the data provider (you and me) and the user (tech giants) and this will need to continue despite being constantly eroded by examples of misuse or abuse or hacking. It's clear what the tech giants need to do to keep their supply of data coming in: there is a balance of utility, trust and transparency in the use data, and increased transparency can often compensate for lower levels of trust.

Once we have more publicly available data and higher levels of transparency from the tech giants we can start to exploit all the connections between these data. Rather than just having public transport data, for instance, we can bring in our own personal data that might include our inherent preferences for how we like to travel. (These could be set manually, of course, but better to pull that information from actual, real-time behaviours rather than what we think we liked a year ago.)

This idea of hyper-localised, hyper-personalised information may well be the key to widespread acceptance of the benefits of AI. Some examples can already be seen (e.g. hyper-localised next-hour weather forecasting, hyper-personalised what-to-wear recommendations) but the bringing to together of all the different sources of data to provide really useful information that is relevant to just you at that specific time in that specific location will be, to use a cliché, a game changer.

If there is trust and transparency around the data that consumers find useful, then they are more likely to allow businesses open access to that, therefore increasing the utility even further. (This is a mind-set change from today where consumers know their data is being exploited but don't ask too many questions because they believe the benefits are usually worth it.) And businesses will benefit from this as well, and can use the same 'three-dimensional data' approach to help run their businesses better internally—salespeople can be much better informed, as can couriers, caterers, HR managers, executives and so on.

So, there is a huge dependency on the levels of trust in data usage, which will be driven by the level of transparency that there is in how the data is farmed and used. Other areas that might delay the date at which AI reaches the plateau of productivity could be a general backlash against the technology, especially as the amount and types of jobs begin to be materially affected.

Right now, though, one of the biggest challenges to the widespread adoption of AI is actually a shortage of skilled and experienced people. Data scientists and AI developers will become superstars (with superstar salaries) as the tech giants and AI organisations, such as OpenAI, fight over these scarce resources. The pipeline for supplying skilled people, which includes the education systems of the world, will take time to adapt, with quick fixes such as online training courses taking the immediate strain.

People will also make a difference on the demand side as well. The Millennials, as we all know, have grown up with many of these technologies and are both familiar and dependent on them. They will absorb new AI applications without thinking and will have fewer issues with some of the ethical challenges I have described (although this is clearly a generalisation—you only have to read the interview with Daniel Hulme in this book to realise that there are people who care deeply about the dangers of AI and how it can benefit the world).

Customers for AI will therefore have a much bigger impact on the adoption of the technology. It is fair to say that AI is currently a vendor-driven market—the innovations and ideas in the market are mainly borne out of the fact that it can be done, rather than it needs to be done. This balance will slowly shift over the coming years as people became more familiar with the capabilities of AI and there are more examples of its use in society and in the workplace.

Things like speech recognition, and particularly voice authentication, have already become part of some people's normal lives as they become familiar with, and then maybe dependent on, their Google Home, Amazon Alexa or Apple HomePod devices. Once this technology enters the business environment, in, say, an HR Helpdesk, we will start to see the beginnings of AI becoming 'business as usual'.

The 'fully connected home' dream that AI product vendors try and sell currently falls far short of expectations—there are lots of different products and standards out there which rarely work together and can be difficult to set up—but one can see signs that the seamless and effortless use of AI to help run our homes is on the horizon. So when people go to work, they will expect the same sort of experience, and this will drive further developments, and the wider acceptance, of AI in business.

One of the most interesting developments I am starting to see is the creation of 'vertical' AI solutions, where a number of different AI (and associated technology) capabilities have been brought together to solve a particular challenge in a particular domain, such as financial services, or HR, or healthcare. Many of the AI 'solutions' being developed right now fall short of being complete solutions—they just look at a very specific capability or service—and these will quickly become commoditised or embedded into existing enterprise systems. 'Full stack' vertical solutions, however, will have AI at their core, and will focus on high-level customer requirements that can be better met using AI, or will be used to discover new requirements that can be satisfied only by using AI. These solutions will take a much more holistic approach and draw upon subject matter expertise just as much as they do technical know-how. In my mind, once we start to see more and more of these types of solutions, the quicker AI will achieve business-as-usual status.

Another bellwether for AI's acceptance onto the plateau of productivity will be how business's organisational structure changes to accommodate AI. There are already a number of enterprises that have created Automation (and AI) Centres of Excellence (I discussed these in Chap. 9) and these will become much more commonplace in the next few years. Full acceptance though will come when enterprises no longer need these CoEs—when AI is seen as the normal way to do business. By this point we won't have AI specialists—everyone will be a generalist with AI skills. Just as in the 1970s there were people who had jobs 'working the computer' and now everyone just 'works with computers', so the same will be true for AI.

There is a slightly tongue-in-cheek definition of AI which says it is 'anything that is 20 years in the future', but there is probably some truth in that. The AI capability that we have now and use every day would have seemed like black magic 20 years ago, yet we don't really consider it to be AI anymore. We are always looking forward for the new shiny thing that we can label AI. So it is almost a self-defeating exercise to try and predict when AI will become business as usual. I have, though, tried in this section to identify those things that will signal when AI (as we consider it today) has reached the plateau of productivity. But if your business is waiting until then before looking into or adopting AI then you will be too late. The time to start your AI journey is now: to future-proof your business so that you can defend against, and prepare for, the new AI-driven business models and opportunities that will inevitably come.

Future-Proofing Your Business

In the very first chapter of this book I implored you to 'not believe the hype'. Over the subsequent nine chapters I have hopefully provided you with a practical guide on how to approach AI so that you are as informed and prepared as you can be. I also hope that you are now as excited as I am about the benefits that AI can bring to business and society, and that you can see how AI could be beneficial to your company.

But the most important thing I want you to have taken away from this is that every company, every senior executive, needs to be thinking about AI now, today. The development of this technology is already moving apace but it will only get faster. Those who wait will be left behind. Now is the time to start future-proofing your business to capture the full value from AI.

There are three broad phases to this: understand the art of the possible, develop an AI strategy and start building stuff.

Understanding the art of the possible is the first stage in future-proofing your business. By reading this book you will have taken an important step forward, but you should also be reading many of the very decent books and resources that are also available, many of them for free on the internet. There are, of course, significantly more poor resources on the internet than there are decent ones, but sites such as *Wired, Quartz, Aeon, Disruption, CognitionX, Neurons, MIT Technology Review, The Economist, The Guardian* and *The New York Times* make very good starting points. You should also consider attending conferences (there are now a plethora of automation and AI conferences available, some of which you may find me speaking at) and also signing up for AI masterclasses and/or boot camps.

The 'art of the possible' should, of course, be grounded with some level of realism. Therefore, it is important to understand what other enterprises are doing, particularly those in the same sector or with the same challenges. Advisors can be useful for this as they (we) tend to have further insights than those that are available publicly. Advisors will also be able to challenge you on what can be done and to help you 'open your mind' to all the available opportunities.

Some useful thought experiments that can be done at this stage include imagining what a company like yours might look like in 10 or 20 years' time, or to try and imagine how you would build a company with the same business model as yours from a blank sheet of paper—what are the core things you would need, what could you discard and what could be delivered through automation? Holding 'Innovation Days' can also work—these would bring together stakeholders from the business to listen to the art of the possible (from experienced third parties) and have demonstrations from AI vendors of their solutions. This should then spark new ideas that can be fed into the AI Strategy.

Following on from the initial thinking, the next phase will all be about **building the AI Strategy**. A good chunk of this book describes this process in detail, so I won't repeat myself here. Suffice it say that the strategy should be built around identifying and addressing existing challenges, as well as identifying new opportunities that can be addressed by AI.

Creating and delivering innovation within any business is never an easy ride. It takes time and effort, with planned investments and a clear mandate. For AI, you could create a small working group with that specific focus, or you could sit on the back of any innovation capability you may already have in-house. The risk with the latter of these is that you may end up with the 'lowest common denominator' and miss out on some of the bigger AI opportunities. The advantage is that you can call upon additional resources and other technologies that may be instrumental to your solutions.

The important thing to remember about any program to implement innovation, but particularly with AI, is that it will require a cultural shift. People will need to think very differently, and they will be challenged about many of their core beliefs of how a business should be run. Be prepared for some heavy discussions.

The final phase to start future-proofing your business is to **build some AI capability**. Based on your AI Strategy, building some prototypes will be the first real test for AI, and the first time that many people will have seen AI in action in your business. This act of disruption will have a profound affect and will be a great catalyst to start new conversations across the business. Ensure that you make the most of this opportunity to showcase AI and bring the rest of the business along with you.

You should always be ready for new developments in the technology. AI is always getting better—advances that we weren't expecting for decades are suddenly with us, and you need to be informed and ready to bring those into your business if they are relevant. Having an AI, or automation, CoE, perhaps backed up with external advice, will ensure that the market is being constantly scanned and all potential opportunities are captured as soon as possible.

Some businesses, once they have some prototypes under their belt, take the disruptive power of AI to the next level and build parallel digital businesses. These in effect compete with the legacy business, and are therefore not encumbered by existing systems or culture. Even if you do not create a new business unit, this healthy competitive mind-set can be useful to bring out the full potential of the technology.

Final Thoughts

This has been a fascinating book to write, and I hope that you have, at least, found it an interesting and thought-provoking read. I'd like to think that you are now much better prepared and informed to start your AI journey with your business.

That journey won't be an easy one—disrupting any business with new technology has more than its fair share of challenges—but the rewards should certainly be worth it. Be very aware that all your competitors will be considering how AI can help their businesses, but the advantage will be to those who actually start doing something about it now. As I said in the previous sections, if you are waiting for a mature and stable AI market then you will be too late.

Thank you for spending your precious time in reading my book. If you need any further help or advice, then you know where to find me. But for now, put down the book, and go start your AI journey.

Index

© The Author(s) 2018
A. Burgess, *The Executive Guide to Artificial Intelligence*,
https://doi.org/10.1007/978-3-319-63820-1